African Agricultural R&D in the New Millennium

Progress for Some, Challenges for Many

Nienke Beintema and Gert-Jan Stads

International Food Policy Research Institute • Washington, DC
Agricultural Science and Technology Indicators • Rome, Italy

February 2011

ABOUT IFPRI

The International Food Policy Research Institute (IFPRI) was established in 1975. IFPRI is one of 15 agricultural research centers that receives its principal funding from governments, private foundations, and international and regional organizations, most of which are members of the Consultative Group on International Agricultural Research.

ABOUT ASTI

The Agricultural Science and Technology Indicators (ASTI) initiative compiles, processes, and disseminates data on institutional developments and investments in worldwide agricultural R&D, and analyzes and reports on these trends. Tracking these developments in ways that facilitate meaningful comparisons among different countries, types of agencies, and points in time is critical for keeping policymakers abreast of science policy issues pertaining to agriculture. The main objective of the ASTI initiative is to assist policymakers and donors in making better informed decisions about the funding and operation of public and private agricultural science and technology agencies by making available internationally comparable information on agricultural research investments and institutional changes. Better-informed decisions will improve the efficiency and impact of agricultural R&D systems and ultimately enhance productivity growth of the agriculture sector.

The ASTI initiative comprises a network of national, regional, and international agricultural R&D agencies and is managed by the International Food Policy Research Institute (IFPRI), a research center of the Consultative Group on International Agricultural Research. ASTI data and associated reports are made freely available for research policy formulation and priority-setting purposes (http://www.asti.cgiar.org).

Cover illustration and design: Joan Stephens, JKS Design.

DOI: http://dx.doi.org/10.2499/9780896295438

ISBN 10-digit: 0-89629-543-5
ISBN 13-digit: 978-0-89629-543-8

Contents

Tables and Figures

Boxes

Acronyms

AgGDP	Agricultural gross domestic product
AGRA	Alliance for a Green Revolution in Africa
ARC	Agricultural Research Council [South Africa]
ARCN	Agricultural Research Council of Nigeria
ASARECA	Association for Strengthening Agricultural Research in East and Central Africa
ASTI	Agricultural Science and Technology Indicators
AU	African Union
AWARD	Agricultural Women in Agricultural Research and Development
CAADP	Comprehensive Africa Agriculture Development Programme
CCARDESA	Centre for Coordination of Agricultural Research and Development in Southern Africa
CGIAR	Consultative Group on International Agricultural Research
CNRA	National Agricultural Research Center [Côte d'Ivoire]
CORAF/WECARD	West and Central African Council for Agricultural Research and Development
CSIR	Council for Scientific and Industrial Research [Ghana]
DRD	Department of Research and Development [Tanzania]
EAAPP	East Africa Agricultural Productivity Program [World Bank]
EIAR	Ethiopian Institute of Agricultural Research
FARA	Forum for Agricultural Research in Africa
FIRCA	Inter-Professional Fund for Agricultural Research and Extension [Côte d'Ivoire]
FTE(s)	full-time equivalent(s)
GDP	gross domestic product
IAC	InterAcademy Council
IFPRI	International Food Policy Research Institute
INRAN	National Agricultural Research Institute of Niger
ISRA	Senegalese Agricultural Research Institute
KARI	Kenya Agricultural Research Institute
MSIRI	Mauritius Sugar Industry Research Institute
NARIs	national agricultural research institutes [Nigeria]
NARO	National Agricultural Research Organization [Uganda]
NEPAD	New Partnership for Africa's Development
PNASA	National Agricultural Services Support Project [Côte d'Ivoire]
PPP	purchasing power parity
R&D	research and development
RARIs	regional agricultural research institutes [Ethiopia]
S&T	science and technology
SADC	Southern African Development Community
SSA	Sub-Saharan Africa
WAAPP	West Africa Agricultural Productivity Program [World Bank]

Acknowledgments

This report was developed with funding from the Bill & Melinda Gates Foundation and additional support from the International Food Policy Research Institute. The authors thank their country collaborators and the more than 370 agricultural research agencies that participated in national survey rounds; without their commitment, this report would not have been possible. The authors also thank Kathleen Flaherty and Michael Rahija for their excellent help in preparing this publication and the two anonymous reviewers for reviewing and commenting on a draft version.

Partners and Contributors

IFPRI gratefully acknowledges the generous unrestricted funding from Australia, Canada, China, Denmark, Finland, France, Germany, India, Ireland, Italy, Japan, the Netherlands, Norway, the Philippines, South Africa, Sweden, Switzerland, the United Kingdom, the United States, and the World Bank.

Executive Summary

After a decade of stagnation during the 1990s, investments and human resource capacity in public agricultural research and development (R&D) averaged more than 20 percent growth in Sub-Saharan Africa (SSA) during 2001–2008. In 2008, the region spent $1.7 billion on agricultural R&D (in 2005 purchasing power parity dollars)—or $0.8 billion (in 2005 constant US dollars)—and employed more than 12,000 full-time equivalent (FTE) agricultural researchers. Most of this growth, however, occurred in only a handful of countries and was largely the result of increased government commitments to augment incommensurately low salary levels and to rehabilitate neglected infrastructure, often after years of underinvestment. Many countries—particularly those in francophone West Africa, which are threatened by extremely fragile funding systems—face fundamental capacity and investment challenges. National investment levels in such countries have fallen so low as to leave them dangerously dependent on often volatile, external funding sources. Despite the overall capacity growth recorded, average qualification levels have deteriorated in a number of countries. Some reported large influxes of BSc-qualified scientists, often in response to prolonged recruitment restrictions, further straining already inadequate training opportunities and far exceeding the capacity for appropriate oversight and mentorship by senior researchers, given years of nonreplacement of retiring and departing scientists.

Notwithstanding the challenges facing many countries, renewed commitment to agricultural R&D by governments and donors indicates improved prospects for agricultural R&D for a number of African countries. Regional initiatives are also a key factor in increasing research coordination and collaboration and ensuring the prioritization and efficiency of research. Increased and sustained investment from national governments, regional and international organizations, and large donors will go a long way toward stabilizing investment and capacity levels and enabling real progress for agricultural R&D in the region.

Building on the strategic recommendations of various highly influential reports and meetings, and taking into account the various investment and capacity challenges outlined in this report, four key areas with strong implications for policy must be addressed by governments, donors, and other stakeholders: (1) decades of underinvestment in agricultural R&D; (2) excessive volatility in yearly investment levels; (3) existing and imminent challenges in human resource capacity; and (4) the need to maximize regional and subregional cooperation in agricultural R&D.

Introduction

The Rationale for Monitoring the Allocation of Agricultural R&D Resources

Extensive empirical evidence demonstrates that agricultural research and development (R&D) investments have greatly contributed to economic growth, agricultural development, and poverty reduction in developing regions over the past five decades (World Bank 2007a; IAASTD 2008). Effectively disseminated new technologies and crop varieties resulting from R&D investments have enhanced the quantity and quality of agricultural produce, at the same time increasing sustainability, reducing consumer food prices, providing rural producers with access to markets, and improving gender-based allocations and accumulations of physical and human capital within households. Given important challenges, such as rapid population growth, adaptation to climate change, increasing weather variability, water scarcity, and the volatility of prices in global markets, policymakers are increasingly recognizing the value of greater investment in agricultural R&D as an essential element in increasing agricultural productivity in Sub-Saharan Africa (SSA).

The 2003 Maputo Declaration directed all member countries of the African Union (AU) to increase agricultural investments to at least 10 percent of their national budgets. To gauge progress toward this target, the Comprehensive Africa Agriculture Development Programme (CAADP) under the AU's New Partnership for Africa's Development (NEPAD) agreed to monitor agricultural expenditures, setting a 6 percent yearly target for growth in agricultural gross domestic product (AgGDP) in countries where agriculture plays a dominant economic role. During 2000–2008, the region's GDP grew by an average of more than 5 percent per year—more than twice as fast as in the two preceding decades—but AgGDP grew by only 3 percent per year on average (Montpellier Panel 2010).

One of CAADP's four foundational pillars focuses on increasing investments in agricultural research, extension, education, and training as a means of promoting growth in agricultural productivity (NEPAD–CAADP 2010). Moreover, NEPAD's African Ministerial Council on Science and Technology established and adopted a Consolidated Plan of Action for developing regional science and technology (S&T). This plan calls for substantial increases in national R&D budgets, with each country taking concrete measures to allocate at least 1 percent of its GDP to R&D (NEPAD 2006). In order to measure, monitor, and benchmark the inputs, outputs, and performance of agricultural S&T systems at the national and regional levels and to assess progress toward the successful implementation of CAADP targets related to S&T, quantitative data are essential. S&T indicators are an indispensable tool when assessing the contribution of agricultural S&T to agricultural growth and, more generally, to economic growth. They assist research managers and policymakers in formulating policy and making decisions about strategic planning, priority setting, monitoring, and evaluation. They also provide information to governments and others involved in the public debate on the state of agricultural S&T at national, regional, and international levels.

This report assesses long-term trends in investments and human resource capacity in public agricultural R&D in SSA, particularly focusing on developments during 2000–2008. The analysis uses information from a set of country notes prepared by the Agricultural Science and Technology Indicators (ASTI) initiative of the International Food Policy Research Institute (IFPRI), using comprehensive datasets derived from primary surveys conducted during 2009–2010. The sample includes 32 countries that contribute more than 90 percent of the region's agricultural GDP.[1] These datasets have been linked with investment and human resources data collected in the region by ASTI during 2002–2003, as well as with ASTI's global datasets, to provide a wider context for agricultural R&D investment trends in SSA over time and in contrast to other regions.[2] The analysis in this report concludes with suggested future directions needed to address the financial and human capacity challenges that many countries currently face.

The Current Institutional Framework of Agricultural R&D in SSA Countries

The institutional structure of most of the region's agricultural research has remained constant since 2000,[3] but Mozambique and Tanzania are important exceptions. The Agricultural Research Institute of Mozambique was established in 2005 with the objective of centralizing agronomic, veterinary, animal production, and forestry research (Flaherty, Mazuze, and Mahanzule 2010). In contrast, Tanzania reversed its earlier decision to consolidate national research activities under the Department of Research and Development (DRD) and instead created a dedicated agency for R&D activities related to livestock (Flaherty and Lwezaura 2010). R&D systems in some other countries also underwent important structural changes. The National Agricultural Research Organisation (NARO) in Uganda, for instance, was transformed from an agency to a consortium in efforts to improve its response to client needs and its ability to oversee and guide agricultural R&D service provision (Flaherty, Kitone, and Beintema 2010). In Nigeria, the Agricultural Research Council of Nigeria (ARCN) was created to improve coordination and linkages across research agencies and between research providers and clients, and to redress overlaps in mandates within the institutes (Flaherty et al. 2010a).

In most of the smaller countries, agricultural research is undertaken by a handful of government agencies and university faculties; systems in the large countries like Ghana, Kenya, Nigeria, South Africa, and Sudan are, understandably, far more complex. Nevertheless, the majority of SSA countries have a single national agricultural research agency that accounts for the bulk of agricultural R&D capacity and investments. Examples include the National Agricultural Research Institute of Niger (INRAN), the Togolese Agricultural Research Institute, and the National Agricultural Research Institute in Eritrea. In some countries, an umbrella organization like Ghana's Council for Scientific and Industrial Research (CSIR) or South Africa's Agricultural Research Council (ARC) oversees and coordinates the R&D activities of a large number of commodity or thematic centers, whereas in a country like Mauritania, the national crop, livestock, and fisheries research agencies operate independently of each other without a coordinating body. Overall, the government sector still dominates agricultural research in the region, but its relative share has declined over time. In 1991, government agencies employed 82 percent of full-time equivalent (FTE) public agricultural R&D staff in SSA on average, but this share had fallen to 73 percent in 2008 (see Box 1 for an explanation of FTEs).

The absolute number of FTE researchers in the higher education sector for the 32 countries included in this study—hereafter referred to as "the ASTI countries"—more than doubled during 1991–2008, mainly as a result of the establishment of new higher education units involved in agricultural research. Most of these new agencies were located in Nigeria and Sudan. Despite the high and increasing number of higher education agencies conducting agricultural research in SSA, the individual capacity of most of them, in terms of FTE researcher numbers, remains small. During 1991–2008, the higher education sector's share of public agricultural research staff (in FTEs) grew from 15 to 24 percent. Although the amount of time staff spend on research has gradually risen over time, in 2008 it still amounted to less than 25 percent of their time on average.

In contrast, the nonprofit sector's share increased marginally during this period, from 2 to 3 percent. The sector's overall growth has been slow compared with the government and higher education sectors. Most nonprofit institutions in SSA are linked to producer organizations and receive most of their funding via levies on production or exports. Although other forms of nonprofit institutions exist in a number of countries, including Benin, Madagascar, and Togo, they play a limited role in agricultural research.

Little information could be obtained on capacity or expenditure trends in private agricultural R&D in SSA.[4] Most private for-profit companies still outsource their research to government agencies or universities, or they import technologies from abroad. Only a limited number of private companies operate their own research programs, and the companies that do so often employ only a handful of researchers. Privately conducted research is estimated to have represented only 2 percent of all public and private investment in agricultural R&D in 2000, and two-thirds of those investments were made in just one country: South Africa (Beintema and Stads 2006).

It is important to note that since 2000 national-level agricultural R&D in SSA has become increasingly interlinked within the region. The Forum for Agricultural Research in Africa (FARA); the Association for Strengthening Agricultural Research in Eastern and Central Africa (ASARECA); the West and Central African for Agricultural Research and Development (CORAF/WECARD); and the Food, Agriculture, and Natural Resources Directorate of the Southern African Development Community (SADC)[5] have all made considerable progress in coordinating agricultural research activities in their member countries through the establishment of various research networks. These networks have proved to be a successful method of collaboration and information sharing. They allow specialization of particular national agricultural research systems in certain fields and have proved to be particularly beneficial for small countries lacking a critical mass of agricultural R&D

Box 1—Measuring agricultural R&D resources

Purchasing power parities (PPPs) as the preferred measure of R&D investments

Comparing R&D data is a highly complex process due to important differences in price levels across countries. The largest components of a country's agricultural R&D expenditures are staff salaries and local operating costs, rather than capital investments that are traded internationally. For example, the wages of a field laborer or lab assistant at a research facility are much lower in Kenya than in any European country; locally made office furniture in Senegal is considerably cheaper than a similar set of furniture bought in the United States.

Standard market exchange rates are the logical choice for conversions when measuring financial flows across countries. However, they are far from perfect currency converters for comparing economic data. At present, the preferred conversion method for calculating the relative size of economies or other economic data, such as agricultural R&D spending, is the purchasing power parity index. PPPs measure the relative purchasing power of currencies across countries by eliminating national differences in pricing levels for a wide range of goods and services. They are also used to convert current GDP prices in individual countries to a common currency. In addition, PPPs are relatively stable over time, whereas exchange rates fluctuate considerably (for example, the fluctuations in the US dollar–euro rates of recent years).

The concept of full-time equivalent (FTE) researchers

ASTI bases its calculations of human resource and financial data on full-time equivalent staffing, or FTEs, which take into account the proportion of time researchers spend on R&D activities. University staff members, for example, spend the bulk of their time on nonresearch-related activities, such as teaching, administration, and student supervision, which need to be excluded from research-related resource calculations. As a result, four faculty members estimated to spend 25 percent of their time on research would individually represent 0.25 FTEs and collectively be counted as one FTE.

Sources: Beintema and Stads (2008a and forthcoming) and ASTI's website (www.asti.cgiar.org/methodology).

staff. Nonetheless, reaching agreement on regional priorities has often been difficult as countries continue to pursue self-sufficiency in fields of agricultural R&D in which they are weak (IAC 2004).

Aside from regional networks, most of the centers under the Consultative Group on International Agricultural Research (CGIAR) have offices in Africa, often with considerable research facilities and staffing. These centers are a key source of agricultural innovation for many countries, providing new crop varieties that are subsequently tested by national agricultural R&D agencies under local conditions. Several other international and regional organizations have a presence and conduct agricultural research in SSA, including the Center for International Cooperation and Agricultural Research for Development (France), the Institute for Research for Development (France), and the World Vegetable Center. Staff and expenditure levels at these international centers are excluded from the analysis of this report because its focus is on national level investments and capacities.

Overview

Investment and Staffing Levels across Countries

Absolute levels of public agricultural R&D spending and staffing varied considerably across the 32 ASTI countries (Table 1). In 2008, Nigeria, South Africa, and Kenya invested $404 million, $272 million, and $171 million, respectively, on agricultural R&D, whereas a further 11 countries spent less than $10 million each, all measured in inflation-adjusted purchasing power parity (PPP) dollars (see Box 1 for an explanation of PPPs). Some countries reported such low investment levels that they were unlikely to have a sustainable impact on rural development and poverty reduction (Beintema and Stads 2006). The 2008 distribution of research staff by country follows a similar pattern, with Ethiopia, Kenya, Nigeria, and Sudan each employing more than 1,000 FTE researchers, and nine additional countries employing fewer than 100 FTEs each.

Table 1—Absolute levels of agricultural R&D spending and staffing, 2008

Spending levels (2005 PPP dollars)			Staffing levels (full-time equivalents)		
> $50 million	$10 to 50 million	< $10 million	> 500	100 to 500	< 100
Nigeria (404)	Côte d'Ivoire (43)	Burundi (9)	Nigeria (2,062)	Mali (313)	Botswana (97)
South Africa (272)	Senegal (25)	Togo (9)	Ethiopia (1,318)	Uganda (299)	Niger (93)
Kenya (171)	Mali (25)	Zambia (8)	Sudan (1,020)	Mozambique (263)	Mauritania (74)
Ghana (95)	Botswana (23)	Mauritania (6)	Kenya (1,011)	Burkina Faso (240)	Rep. of Congo (71)
Uganda (88)	Mauritius (22)	Niger (6)	South Africa (784)	Guinea (229)	Namibia (70)
Tanzania (77)	Namibia (22)	Sierra Leone (6)	Tanzania (674)	Madagascar (212)	Sierra Leone (67)
Ethiopia (69)	Benin (22)	Rep. of Congo (5)	Ghana (537)	Zambia (209)	Togo (63)
Sudan (51)	Malawi (21)	Guinea (4)		Mauritius (158)	Gabon (61)
	Burkina Faso (19)	Eritrea (3)		Senegal (141)	The Gambia (38)
	Rwanda (18)	The Gambia (3)		Zimbabwe (139)	
	Mozambique (18)	Gabon (2)		Malawi (127)	
	Madagascar (12)			Côte d'Ivoire (123)	
				Eritrea (122)	
				Benin (115)	
				Burundi (107)	
				Rwanda (104)	

Source: Compiled by authors based on country-level ASTI survey data and several secondary resources. (For more information, see individual ASTI Country Notes available at www.asti.cgiar.org.)

Notes: Total 2008 spending and capacity levels are shown in parentheses. Spending data for Zimbabwe were unavailable. Total expenditures for a number of countries differ from those published in ASTI's Country Notes because of recent World Bank revisions to their GDP deflator; for Kenya, Guinea, Madagascar, Mauritius, Mozambique, Nigeria, and Zambia this change was in the order of 2 to 20 percent. Further information on ASTI's data methodology and calculation procedures is provided on ASTI's website (www.asti.cgiar.org/methodology); see also Beintema and Stads (forthcoming).

Analyzing absolute levels of research expenditures explains only so much. Another way of comparing the commitment to public agricultural R&D investments across countries is to measure total public agricultural R&D spending as a percentage of AgGDP. This relative measure indicates the intensity of investment in agricultural research, not just the absolute level of spending. In 2008, SSA invested $0.61 for every $100 of agricultural output on average (Figure 1a), which was below NEPAD's national R&D investment target of at least 1 percent of GDP. Only 8 of 31 ASTI countries (excluding Zimbabwe) for which data were available met this 1 percent target. Burundi, Kenya, Mauritania, Namibia, South Africa, and Uganda recorded 2008 ratios of between 1.2 and 2.0 percent, whereas Mauritius and Botswana recorded particularly high

Figure 1—Relative levels of agricultural R&D spending and staffing, 2008

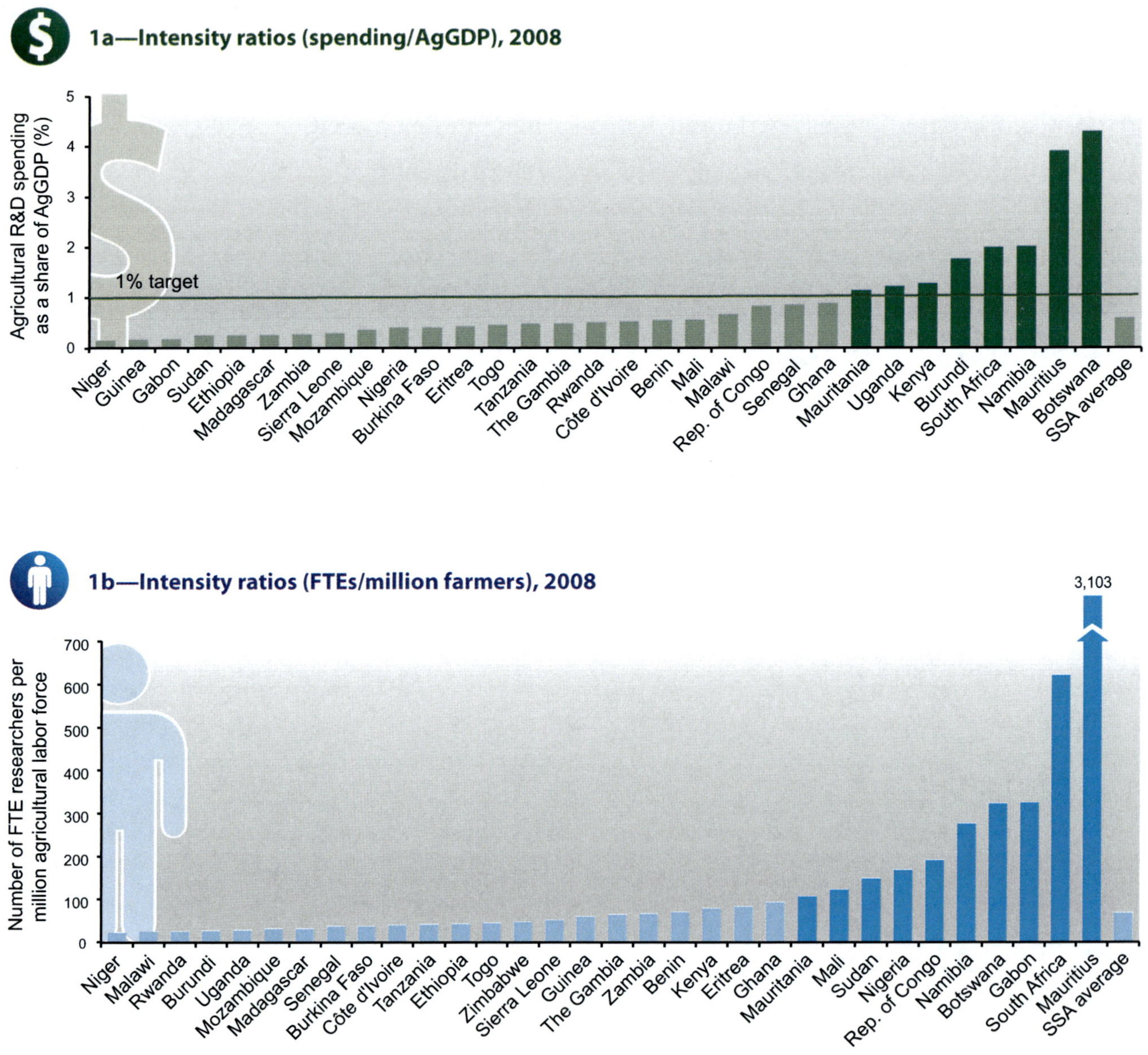

Sources: Compiled by authors based on country-level ASTI survey data and several secondary resources (see individual ASTI Country Notes). AgGDP data are from World Bank 2010; economically active agricultural population data (here labeled as agricultural labor force) are from FAO 2009.
Notes: Total expenditures for Kenya, as well as Benin, Côte d'Ivoire, Eritrea, Guinea, Mauritius, Niger, Sierra Leone, and Uganda (although less pronounced) differ from the total published in the ASTI Country Notes because of recent World Bank revisions to their GDP deflator.

ratios of 3.9 and 5.2 percent, respectively. For Mauritius this reflects the high level of investment in sugarcane research. In contrast, a large number of countries recorded intensity ratios of 0.5 percent or lower.

Gauging researcher numbers or spending levels against total population or economically active agricultural population also facilitates cross-country comparisons (Figure 1b).[6] In 2008, SSA employed 68 FTE researchers per million economically active agricultural population. Again, wide variation occurred across the ASTI countries. Botswana, Gabon, and Namibia employed more than 200 FTE researchers per million economically active agricultural population. A particularly high ratio in Mauritius (3,103 FTEs) again reflects the leading role the country plays in sugarcane research. Gabon, Nigeria, and Sudan had relatively high ratios of research staff compared with their spending intensities.

Although intensity ratios provide useful insights into relative investment and capacity levels across countries, they take into account neither the policy and institutional environment within which agricultural research occurs nor the broader size and structure of a country's agricultural sector and economy. For example, small countries require more human resource and capital investments, because they are unable to benefit from the economies of scale available to larger countries. More important, a high intensity can actually reflect reduced agricultural output rather than higher investment, as is noted for Botswana. Detailed analysis is needed to ensure a clear understanding of the implications of intensity ratios across countries.

Uneven Investment and Staffing Trends over Time

In the late-twentieth century, greater instability was evident in agricultural R&D in SSA compared with other world regions, mainly due to political unrest, social and economic hardship, and institutional vulnerability. Spending levels fluctuated in many countries, and overall growth slowed over time (Beintema and Stads 2006). This trend appears to have reversed, at least in the aggregate for the 2001–08 period. In 2008, public agricultural R&D investments for SSA as a whole—based on data for the 32 ASTI countries and estimates for 14 other, often small countries—totaled $1.7 billion in inflation-adjusted PPP dollars—or $0.8 billion in 2005 constant US dollars. This was almost 20 percent higher than the $1.4 billion (in 2005 PPP dollars)—or $0.6 billion (in 2005 US dollars)—recorded in 2001 and marks a considerable shift away from the slow 1.0 percent yearly growth in agricultural R&D investments recorded during the 1990s (Figure 2 top graph).[7] Overall, 2008 investment levels in SSA were comparable to those in individual countries like Brazil and India with high investment levels (Box 2).

Growth in agricultural R&D capacity was strong in the 1970s and 1980s, at 5.4 and 3.8 percent per year, respectively, but during the 1990s it slowed to a mere 1.3 percent per year. Since the turn of the millennium, growth in researcher numbers has once again accelerated. In some countries, renewed growth was due to the cessation of long-term recruitment bans, whereas in other countries it stemmed from increased involvement in agricultural research by the higher education sector. In 2008, SSA employed 12,120 FTE researchers, compared with 9,824 FTEs in 2001.

The relative growth, in terms of the intensity ratio, has also increased since the turn of the millennium. The aforementioned 2008 investment of $0.61 for every $100 of agricultural output was considerably higher than the average of $0.49 during the late-1990s (Figure 2 bottom data). This was mainly due to the aforementioned low growth in agricultural R&D spending during that decade, along with higher increases in AgGDP. Although SSA's intensity ratio has increased since 2000, it is still below the levels of the late 1980s and early 1990s.

Country-level data reveal that the regionwide spending and capacity increases of roughly 20 percent during 2001–2008 were largely driven by only a handful of countries. More than one-third of the growth in public agricultural R&D spending during this period is attributable to a $110 million increase in spending in Nigeria. Ghana, Sudan, Tanzania, and Uganda also experienced relatively high increases in total spending of between $25 million and $56 million each (in 2005 PPP dollars) (Figure 3a). In contrast, Ethiopia and South Africa experienced notable declines ($28 million and $12 million, respectively).

Figure 2—Trends in total public agricultural R&D spending and staffing, 1971–2008

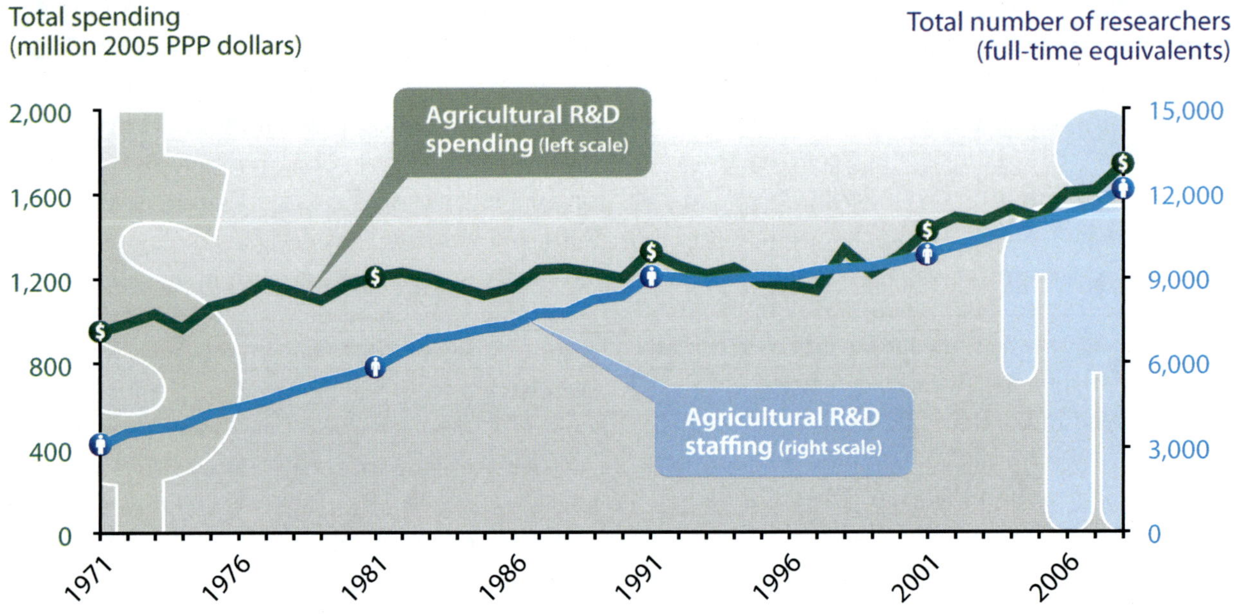

Total spending

1971	**$963 million**
1981	**$1,218 million**
1991	**$1,335 million**
2001	**$1,432 million**
2008	**$1,741 million**

Total FTE researchers

1971	**3,060 FTEs**
1981	**5,819 FTEs**
1991	**9,065 FTEs**
2001	**9,824 FTEs**
2008	**12,120 FTEs**

Yearly rates of spending growth (2005 prices)

1971–1981	**1.7%**
1981–1991	**0.6%**
1991–2001	**1.0%**
2001–2008	**2.4%**

Yearly rates of researcher growth (FTEs)

1971–1981	**5.4%**
1981–1991	**3.8%**
1991–2001	**1.3%**
2001–2008	**2.8%**

Spending per $100 agricultural output (AgGDP)

1971–1975	**$0.63**
1976–1980	**$0.69**
1981–1985	**$0.68**
1986–1990	**$0.58**
1991–1995	**$0.59**
1996–2000	**$0.49**
2001–2005	**$0.59**
2008	**$0.61**

FTE researchers per million agricultural labor force

1971–1975	**na**
1976–1980	**na**
1981–1985	**63 FTEs**
1986–1990	**67 FTEs**
1991–1995	**68 FTEs**
1996–2000	**65 FTEs**
2001–2005	**64 FTEs**
2008	**68 FTEs**

Sources: Compiled by authors from country-level ASTI survey data and several secondary resources (see individual ASTI Country Notes). AgGDP data are from World Bank 2010; economically active agricultural population data (here labeled as agricultural labor force) are from FAO 2009.

Notes: The figure includes 46 SSA countries but excludes Djibouti and Somalia because macroeconomic data were not available. Data on research spending and capacity for the 14 non-ASTI countries (accounting for 10 percent of total agricultural output in SSA) were estimated based on their share of total agricultural output. Yearly growth rates are calculated using the least-squares regression method.

Nigeria was also the main driver of the regional growth in researcher numbers, accounting for 724 of the region's 2,285 increase in FTE researcher numbers during 2001–2008 (Figure 3b). Ethiopia, Kenya, and Sudan reported significant R&D staff increases as well. South Africa experienced the largest decline in public agricultural researcher numbers (140 FTEs), whereas changes in investment and capacity levels in the remaining ASTI countries were less severe in absolute terms during 2001–2008.

Box 2—Sub-Saharan Africa compared with Brazil, China, and India

In 2000 (the latest year for which global data were available), SSA contributed 5 percent of the $25 billion spent on public agricultural R&D globally (in 2005 PPP prices), compared with 7 percent in 1981. This decline resulted in part from relatively low yearly spending growth during the 1990s, combined with a very strong increase in public agricultural R&D spending in the Asia-Pacific region, specifically in China and India (Beintema and Stads 2010). No recent information on worldwide public agricultural R&D spending is available, but investments in Brazil, China, and India have continued to rise, so SSA's overall share is unlikely to have increased despite gains since 2000 (Figure 2). Increases in Brazil, China, and India were mostly the result of renewed government commitment to public agricultural R&D rather than increased donor funding, which is low compared with levels in many SSA countries. By way of magnitude, Brazil and India both spent slightly less on public agricultural R&D than SSA as a whole. China's spending far exceeds any other country's and in 2007 totaled $4.3 billion (in 2005 PPP prices).

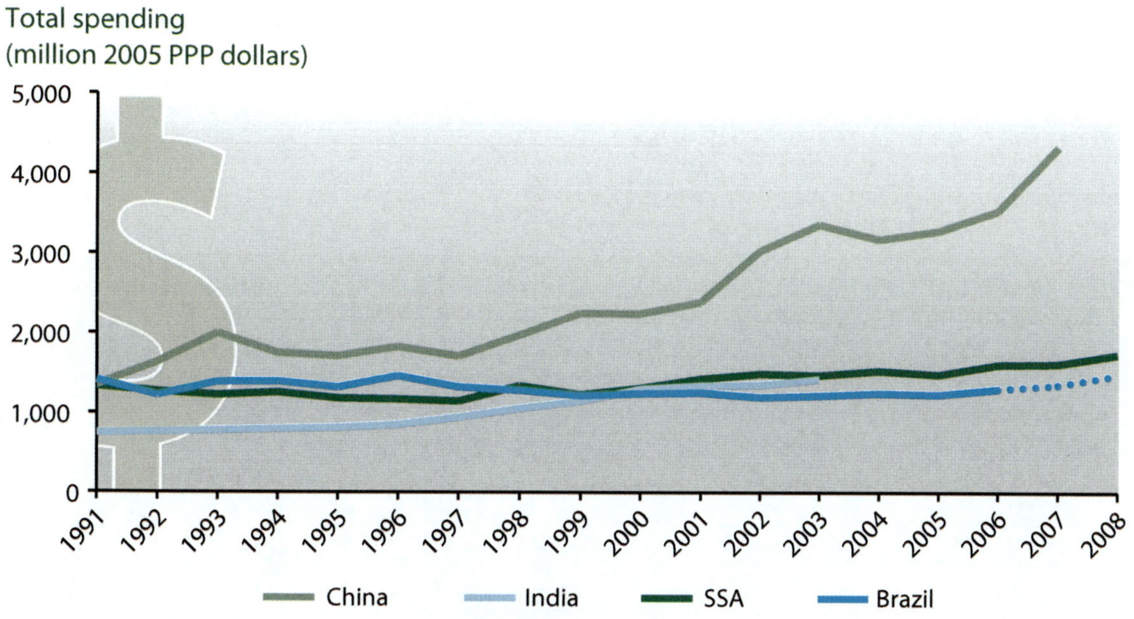

FTE researchers

China (2007)	**na**
India (2003)	**16,703 FTEs**
SSA (2008)	**12,120 FTEs**
Brazil (2006)	**5,376 FTEs**

Sources: Sub-Saharan Africa data are from Figure 1; Brazil, China, and India data are from Beintema and Stads (2010).

Figure 3—Main drivers of growth in agricultural R&D spending and staffing, 2001–2008

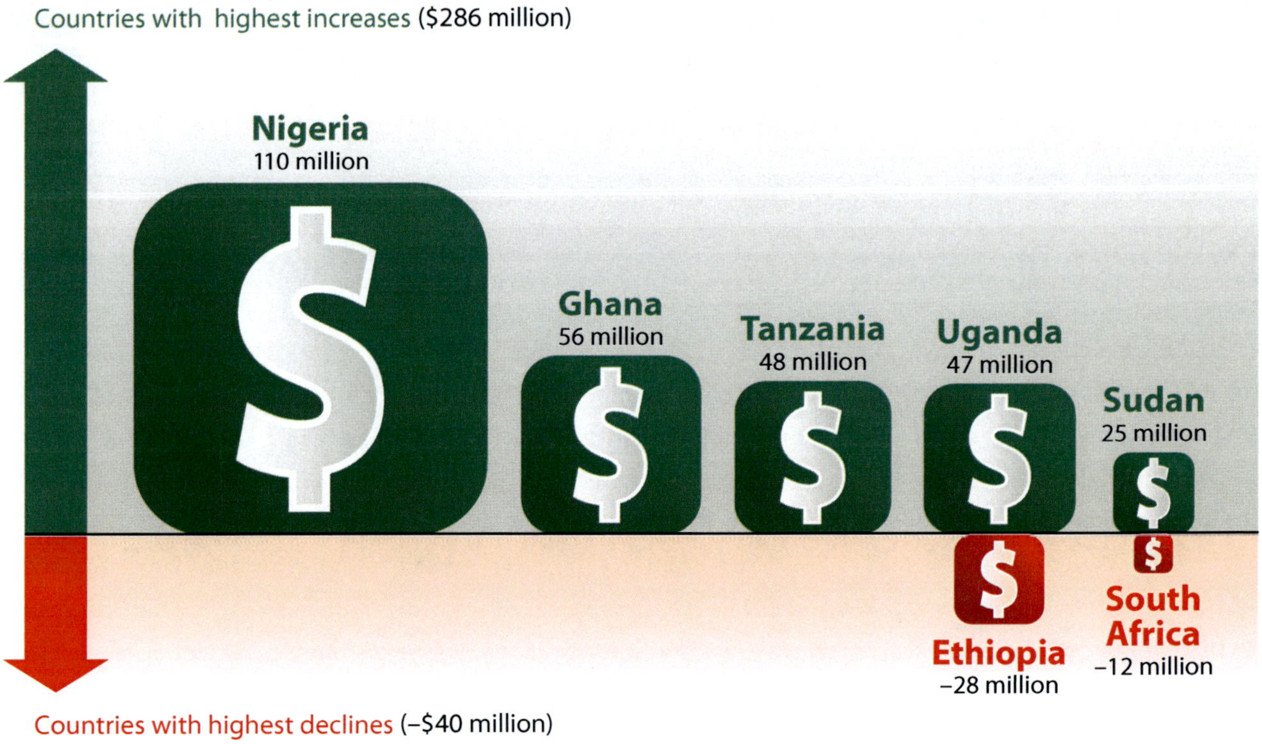

3b—Overall capacity increase in SSA: 2,285 FTE researchers

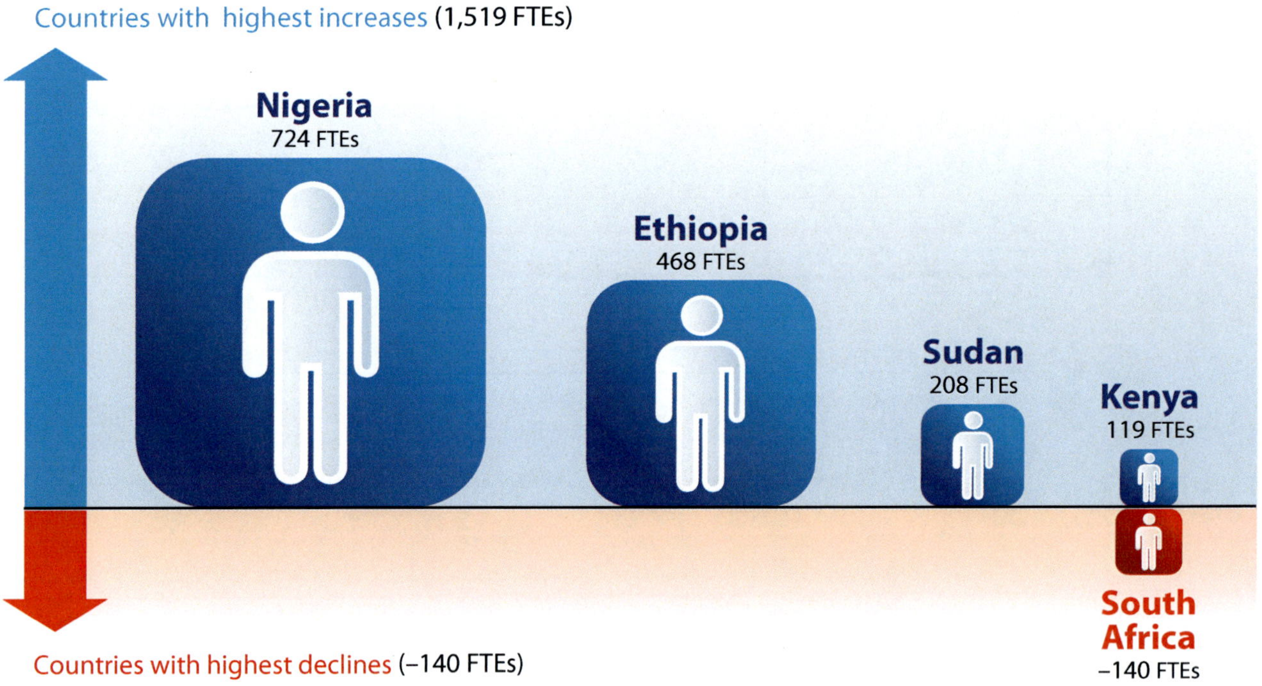

Sources: Compiled by authors based on country-level ASTI survey data and a few secondary resources (see individual ASTI Country Notes).

The Drivers of Regional Growth

Africa's "Big Eight"

Unsurprisingly, the countries identified as the main drivers of regional growth in agricultural R&D spending and capacity were those with the largest absolute spending and capacity, as indicated in Table 1. These "Big Eight"[8]—Nigeria, South Africa, Kenya, Ghana, Uganda, Tanzania, Ethiopia, and Sudan—accounted for 70 percent of regional public agricultural R&D spending and 64 percent of all researchers in 2008 (Table 2). This is a considerable increase above 1991 levels, when the corresponding shares were 53 and 55 percent, respectively. Because of the relative size of these countries, it is not surprising that developments in capacity or spending, whether positive or negative, have such a significant impact on regional trends, so these countries warrant individual attention.

Nigeria. Nigeria has the largest agricultural research system in SSA in terms of investments, capacity, and the number of government and higher education agencies. After a period of stagnation during the late 1980s and early 1990s, public agricultural R&D more than doubled during 2000–2008. This was the result of renewed government commitment to agricultural R&D that increased salary levels and substantial investments in research infrastructure and equipment. In addition, Nigeria's agricultural research staffing levels grew steadily during the same period. Notably, the role of the higher education sector in agricultural research increased during this time. Although growth in Nigerian agricultural R&D spending and capacity was striking, growth only served to offset years of severe underinvestment in the 1990s (Flaherty et al. 2010a).

South Africa. South Africa has one of the most well-established and well-funded research systems in SSA. Nevertheless, the country's yearly public agricultural R&D expenditures varied significantly during 2000–2008, largely due to fluctuations in government funding

Table 2—Big Eight shares of regional public agricultural R&D spending and staffing, 1991 and 2008

Country	Share of regional spending (%)		Share of regional staffing (%)	
	1991	2008	1991	2008
Nigeria	9.8	23.2	11.8	17.0
South Africa	19.3	15.6	10.7	6.5
Kenya	10.4	9.8	10.7	8.4
Ghana	2.3	5.5	3.6	4.4
Uganda	1.7	5.0	2.6	2.5
Tanzania	2.0	4.4	5.9	5.6
Ethiopia	2.1	3.9	4.0	10.9
Sudan	5.7	3.0	5.5	8.4
Subtotal (8)	**53.3**	**70.4**	**54.8**	**63.7**

Source: Compiled by authors based on country-level ASTI survey data and several secondary resources (see individual ASTI Country Notes).

to ARC, the country's main agricultural R&D agency. Agricultural researcher numbers declined by one-third between 1997 and 2004, and capacity increased only slightly thereafter (Flaherty, Liebenberg, and Kirsten 2010).

Kenya. Public agricultural R&D in Kenya is well funded and well staffed compared with many other countries in SSA. Yearly agricultural R&D investments in Kenya did vary, however, during 2000–2008, reflecting fluctuations in donor funding and, to a lesser extent, national government funding to the Kenya Agricultural Research Institute (KARI). In contrast, agricultural research capacity remained stable. Following a period of decline in the late 1990s, Kenya's number of agricultural researchers increased as a combined result of the merger of two institutes with KARI and the relaxation of a government recruitment freeze (Flaherty et al. 2010b).

Ghana. After a period of relative stagnation in the 1990s, agricultural R&D spending in Ghana more than doubled during 2000–2008, largely due to increased investments by CSIR agencies and increased revenues at the Cocoa Research Institute of Ghana resulting from growth in cocoa production. Agricultural research staffing also grew steadily throughout this period, albeit at a much slower rate than expenditures. However, some of the country's agencies recorded uneven growth, and others reported a decline in researcher numbers between 2000 and 2008 (Flaherty, Essegbey, and Asare 2010).

Uganda. Investments in public agricultural R&D in Uganda quadrupled during 2000–2008, primarily as a result of increased donor funding and development bank loans, along with growth in government funding to NARO after 2005. Human resource capacity began to rebound in the mid-2000s after a period of falling staffing levels due to losses at NARO in response to low salaries and a hiring freeze (Flaherty, Kitone, and Beintema 2010).

Tanzania. Tanzania's agricultural R&D system has traditionally been highly dependent on donor funding, which has fluctuated considerably. Since 2005, the government has stepped up its support to agricultural research. Total agricultural research capacity has grown slightly in recent years, with most of the growth taking place in the higher education sector (Flaherty and Lwezaura 2010).

Ethiopia. Agricultural research spending doubled between 1993 and 2000, and then doubled again during 2000–01 following increased government and donor support. By 2008, however, expenditures at the main agricultural research agency, the Ethiopian Institute of Agricultural Research (EIAR), had reverted to 2000 levels. Agricultural research staffing at the regional agricultural research institutes (RARIs) and universities grew significantly during 2000–08, such that by 2008 the combined capacity of the seven RARIs was higher than that of EIAR in terms of staffing levels, but not in terms of postgraduate qualifications (Flaherty, Kelemework, and Kelemu 2010).

Sudan. During the 1990s agricultural R&D spending in Sudan declined rapidly due to general neglect of the agricultural sector in favor of large-scale oil production. This trend was reversed more recently with increased national government support for agricultural development, enabling greater R&D investment and hence recovery of lost ground. Agricultural research capacity also rose considerably after 2000 (Stads and El-Siddig 2010).

Increased Spending, But on What?

As established, the overall increase in agricultural R&D investments in SSA is driven by a handful of countries, but the underlying cost-category breakdown reveals that different factors drove growth. The rapid increase in Ghanaian agricultural R&D spending, for instance, was driven almost entirely by increased salary expenditure at CSIR rather than expanded research activities or greater investment in equipment or infrastructure (Figure 4A). The unprecedented increase in expenditure on salaries needs to be understood in the context of years of underfunding, during which salary levels became increasingly incommensurate and uncompetitive (Flaherty, Essegbey, and Asare 2010). At DRD in Tanzania, on the other hand, spending on salaries has remained relatively stable over time (Figure 4B). Prior to 2005, DRD was highly dependent on World Bank funding and capital investments were high, but spending plummeted once the World Bank project ended. Thereafter, the Tanzanian government increased its commitment to agriculture and agricultural research over time, thereby allowing greater expenditure on research activities and equipment and infrastructure (Flaherty and Lwezaura 2010). Similarly,

Figure 4—Spending by cost category for the main agricultural R&D agencies in selected Big Eight countries, 2001–2008

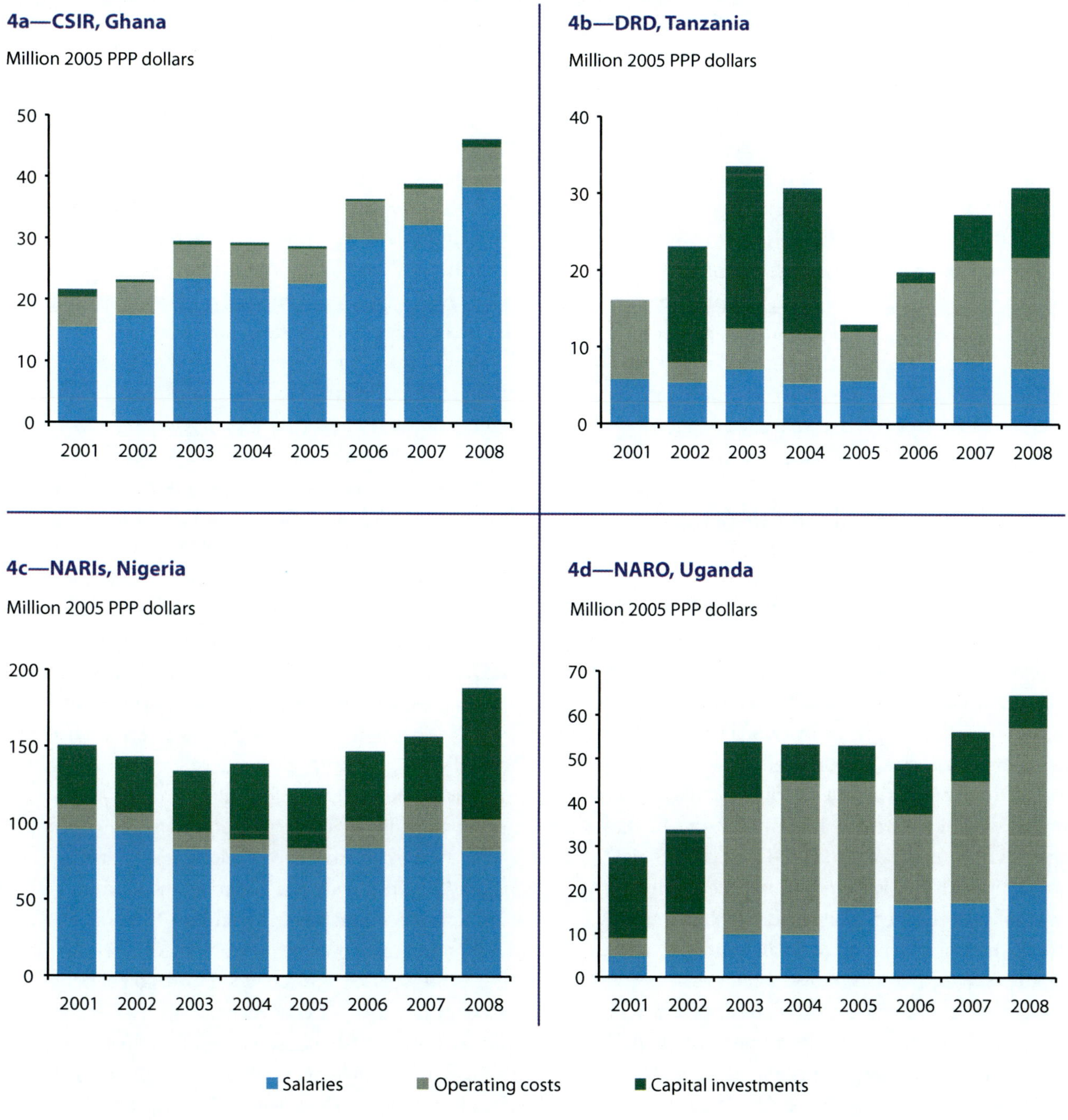

Source: Compiled by authors based on country-level ASTI survey data.

Notes: Ghana includes CRI, FORIG, FRI, OPRI, PGRRI, SARI, SRI, and WRI; Nigeria includes CRIN, IAR, IAR&T, LCRI, NAERLS, NAPRI, NCRI, NIFFR, NIFOR, NIHORT, NIOMR, NRCRI, NSPRI, NVRI, and RRIN (see individual Country Notes for full details of agency names).

the government of Nigeria significantly increased its funding to the national agricultural research institutes (NARIs) and other government agencies from the late 1990s, enabling the purchase of equipment and the rehabilitation of facilities (Figure 4C). Nonetheless, despite these remarkable increases, investment levels remain below those required to restore facilities to prior levels and to sustain the country's agricultural research needs (Flaherty et al. 2010a). In Uganda, sizable World Bank support and, more recently, increased support from the Ugandan government have enabled the country's NARO to invest in institutional development, research programs, rehabilitation of research infrastructure, and postgraduate training (Figure 4D). In 2005, the Ugandan government approved a much-needed 100 percent salary increase and thereafter a yearly increase of 10 percent. Understandably, this had a significant impact on overall spending levels (Flaherty, Kitone, and Beintema 2010).

Who's Catching Up and Who's Lagging Behind?

Though increases and decreases in the absolute levels of agricultural R&D spending and capacity of the Big Eight overshadow those of many of the smaller countries in SSA, a closer look at relative shifts in investment and capacity levels over time reveals some interesting cross-country differences and challenges. Some of the region's smallest countries have such low and declining levels of investment and human resource capacity that the effectiveness of their national agricultural R&D could be questioned. This also highlights the need for regional initiatives to address the unique needs of small countries and to take advantage of collaborative synergies.

Considerable differences were reported not only in absolute investment levels among the ASTI countries but also—more importantly—in the magnitude of growth over time. During 2001–2008, 13 of 29 countries (excluding Rwanda, Mozambique, and Zimbabwe) experienced negative yearly growth in public agricultural R&D spending, ranging from –1.6 to –12.4 percent per year (Figure 5A), which is sizeable given that spending in SSA as a whole actually increased throughout this period. Of these 13 countries, 7 are francophone countries located in West and Central Africa. With the exception of Gabon and Mali, these countries also experienced negative growth during the 1990s, which is a major area of concern. Falling investments in agricultural R&D in these countries resulted mainly from the completion of large donor-funded projects, often financed through World Bank loans (Burkina Faso, Guinea, Senegal, and Togo). Comparing the 2001–2008 growth rates with those of the 1990s clearly illustrates the volatility of agricultural spending levels for many of the region's countries. Eritrea and Ethiopia, for example, experienced negative growth during 2001–2008 (–12.4 and –4.5 percent per year, respectively) following a decade of particularly high positive growth (32.7 and 11.0 percent, respectively), which is indicative of high dependency on donor funding (see the next section for details).

In contrast, agricultural R&D investments in a number of countries increased substantially after 2000. Eight countries recorded yearly growth rates of more than 6 percent, including four of the Big Eight countries (Ghana, Sudan, Tanzania, and Uganda). Spending in Nigeria, which accounted for more than one-third of the absolute growth in spending during 2001–2008, grew at a comparatively moderate average rate of 3.2 percent per year. For some countries, growth reflected the re-establishment of agricultural R&D systems after periods of political unrest, whereas in others—such as Nigeria, Sudan, Tanzania, and Uganda—growth stemmed from increased national government commitment to agriculture in general and to agricultural R&D in particular.

Growth in agricultural research staffing varied less across countries compared with total spending (Figure 5B). In line with reduced spending levels, a number of francophone countries in West and Central Africa also reported declining capacity during 2001–2008. Gabon, Niger, the Republic of Congo, and Togo recorded yearly declines of –1.8 percent or more. This is a worrisome trend because capacity in these countries was low to begin with. Even more challenging, the pool of well-qualified and experienced researchers in many countries is aging; many will be lost to retirement in the next decade (see "Staffing Trends" for details). Nevertheless, research staffing increased or remained fairly constant in most countries during 2001–2008. Despite strong losses in spending levels in Eritrea and Ethiopia throughout this period, researcher numbers increased by 6.6 and 6.0 percent per year, respectively. In Ethiopia this growth was driven by the development of the RARIs and the higher education sector.

Figure 5—Compound yearly rates of growth across countries, 2001–2008 compared with 1991–2001

5a—Annual growth rates, spending

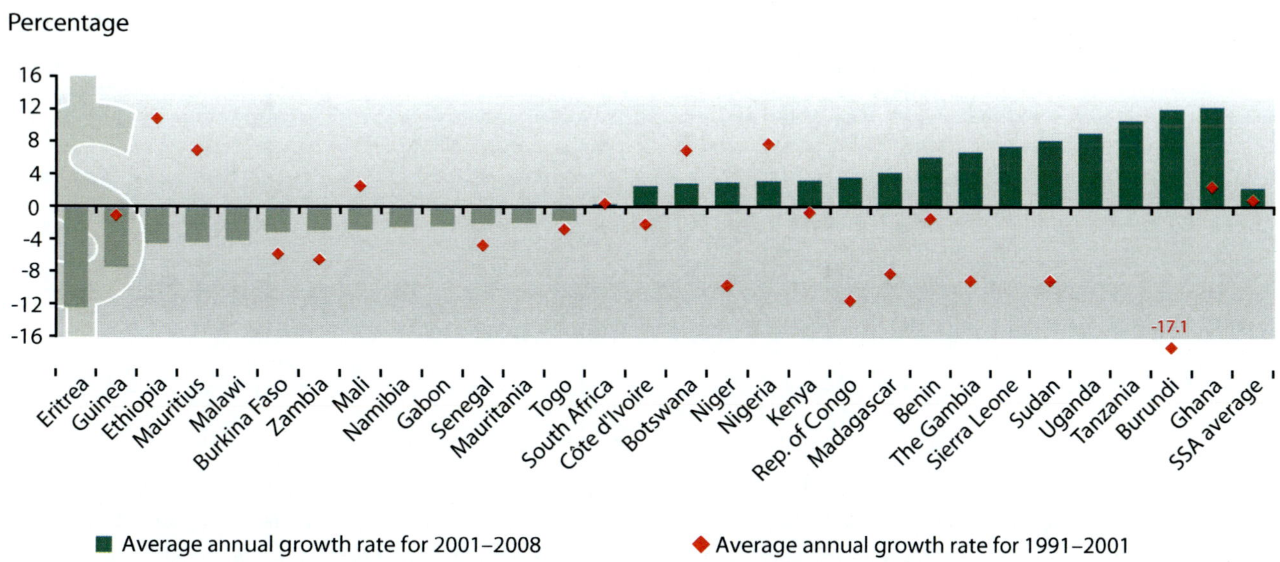

5b—Annual growth rates, researchers

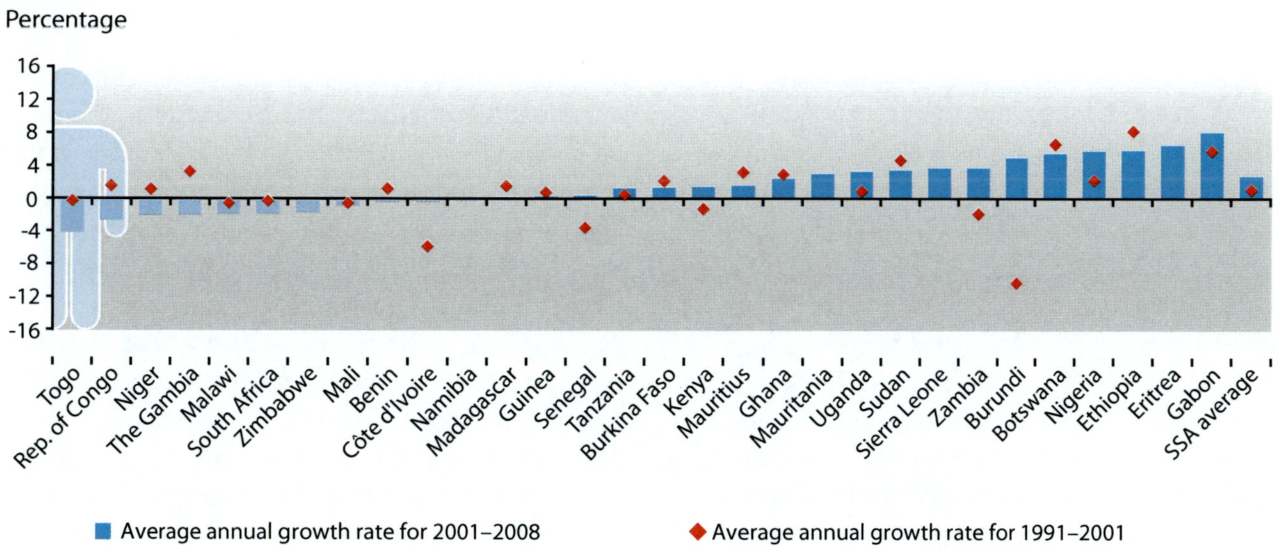

Sources: Compiled by authors based on country-level ASTI survey data and several secondary resources (see individual ASTI Country Notes).

Notes: The figure excludes Mozambique, Rwanda, and Sierra Leone (for both spending and capacity) and Malawi and Zimbabwe (for spending only) because time-series data did not date back to 2001. Growth rates are missing for Eritrea, Mauritania, and Namibia (for spending and capacity) and Tanzania and Uganda (for spending only) due to lack of time-series data for the full period of 1991–2001. Compound yearly growth rates are calculated using the least-squares regression method.

Funding Trends

Who's Footing the Bill?

Funding for African agricultural R&D is derived from a variety of sources, including national governments; donors, development banks, and (sub)regional organizations; producer organizations; the private sector; and internally generated revenues. Given that funding data were not available for all 32 ASTI countries (including some important Big Eight countries), it is not possible to present regionwide funding trends. Evidence presented above, however, indicates that growth in spending in Ghana, Nigeria, Sudan, Tanzania, and Uganda—the main drivers of regional growth—was largely the result of significant injections of government funding.

Unsurprisingly, variation is significant across countries and agencies (Figure 6). The government funds the bulk of agricultural R&D activities in some countries, whereas other countries are extremely dependent on external funding. A number of R&D agencies generate significant revenues by selling goods and services, whereas for others the proceeds of sales are channeled to the national treasury, thereby eliminating the incentive to pursue such endeavors.

In addition to differing across countries at a specific time, sources of funding differ substantially across time. KARI, for example, is a relatively well-funded institute, receiving constant support from the Kenyan government, attracting large sums of donor funding, and generating its own revenues (Figure 7A; Flaherty et al. 2010b). In contrast, the Mauritius Sugar Industry Research Institute (MSIRI), a nonprofit agency, is almost entirely funded through a tax on sugar production (Figure 7B). Given falling world market prices of sugarcane in recent years, overall funding to MSIRI has declined (Rahija, Ramkissoon, and Stads 2010). Côte d'Ivoire's National Agricultural Research Center (CNRA) is funded largely by the private sector (mostly coffee, cocoa, rubber, and oil palm producers) supplemented by limited funding from the national government (Figure 7C). Foreign donors stopped supporting the center shortly after the civil war broke out in 2002, and they have not returned since (Stads and Doumbia 2010). Niger's INRAN is an example of an institute that was extremely dependent on donor and development bank funding during the 1990s, but with the completion of a large World Bank–financed project in 1998, the institute fell into severe financial crisis, and the situation remains precarious (Figure 7D; Stads, Issoufou, and Massou 2010). Like INRAN, many other African institutes have extremely fragile and donor-dependent funding systems.

The higher education sector is excluded from this section because agricultural R&D funding data for this sector are extremely difficult to obtain. Given that teaching is the core business of agricultural faculties, dedicated R&D budgets are rare or ad hoc. Many universities fund R&D activities through public grants, student tuition fees, and internally generated resources. Like the government sector, donor funding plays an important role in many countries. African universities often maintain close linkages with universities in developed countries and benefit from funding as part of joint research projects. Although the role of higher education in agricultural R&D in SSA could increase if it were to receive sustainable research funding, it is unlikely that funding would increase to levels seen in countries like India, Mexico, or the United States. Agricultural higher education agencies in SSA are often fragmented and fall under universities with a broader (nonagricultural) mandate; independent agricultural universities or colleges are scarce. Further, many higher education agencies in SSA are underfunded and understaffed and have therefore limited time available for activities other than education (World Bank 2007c).

Figure 6—Relative shares of funding sources for the main agricultural R&D agencies, 2008

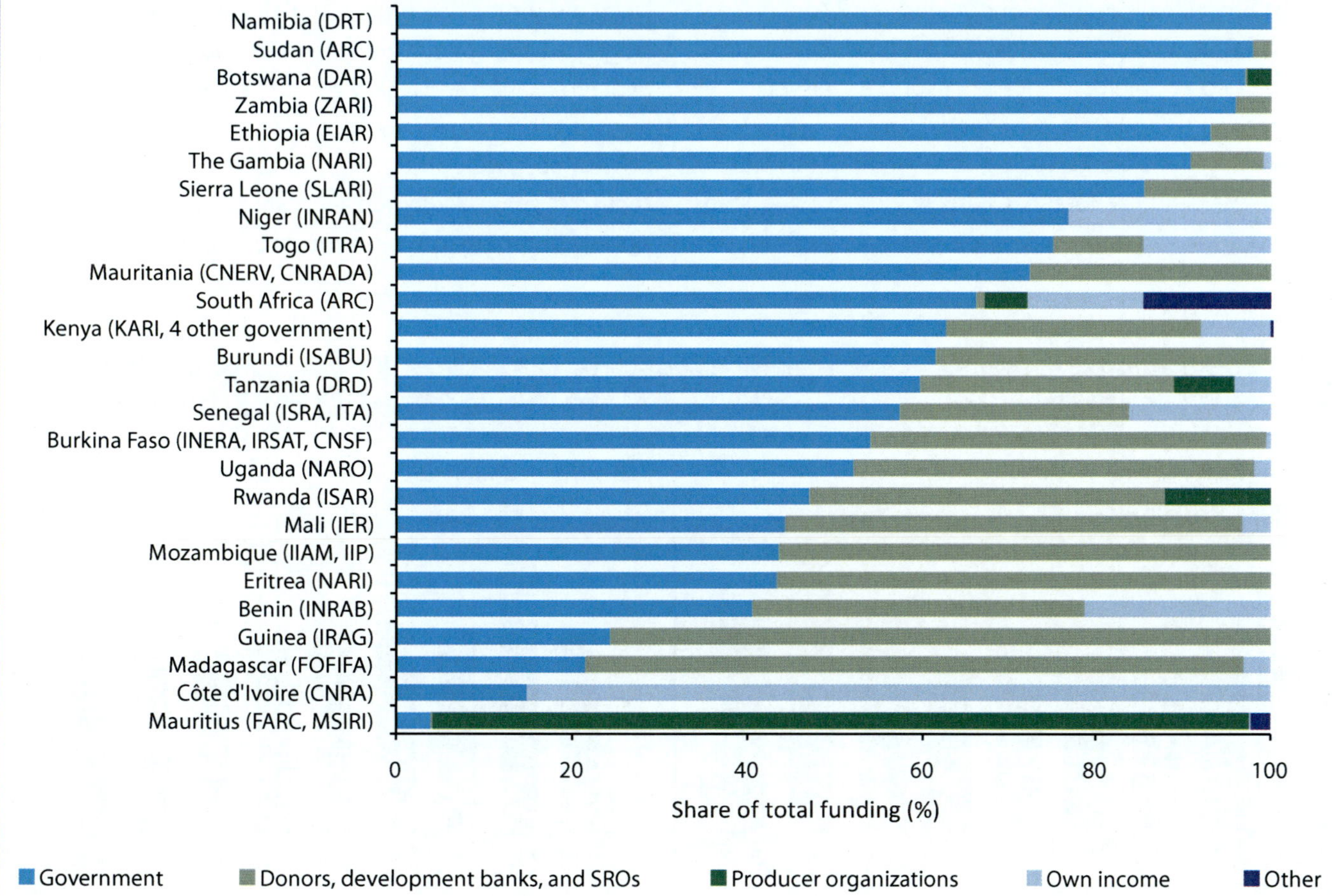

Source: Compiled by authors based on country-level ASTI survey data.

Notes: Gabon, Ghana, Malawi, Nigeria, the Republic of Congo, and Zimbabwe were excluded due to lack of complete data. SROs indicates subregional organizations; "producer organizations" include contributions through export or production levies; "own income" includes sales of goods and services and contractual research performed for public and private agencies. Funding shares for some research agencies fluctuated over time.

Sources and Mechanisms of Funding

National Governments. In southern African countries like Botswana, Namibia, and Zambia, more than 95 percent of agricultural R&D is financed by the national government, whereas in countries like Guinea and Madagascar government funding plays a relatively limited role. For many agencies, fluctuations in government funding can undermine gains achieved by making it difficult for agencies to retain senior staff, attract and train new staff, develop and maintain appropriate facilities, coordinate activities with other agencies, and ultimately sustain viable research programs.

Donors, Development Banks, and (Sub)Regional Organizations. Donors, development banks, and (sub)regional organizations contribute about 3 percent of agricultural R&D funding in Latin America and the Caribbean and in the Asia–Pacific (Beintema and Stads 2008b; Stads and Beintema 2009). Shares of funding in this category are generally much higher in SSA (Figure 6), although they are negligible in many middle-income countries in southern Africa or in countries afflicted by political unrest. Donor funding is provided by:

- multilateral bodies, such as the European Union, CGIAR, and the United Nations;

- bilateral donors, such as foreign governments and private foundations;
- (sub)regional organizations such as FARA, ASARECA, and CORAF/WECARD, which, in turn, are also recipients of donor resources (largely through multidonor trust funds); and
- development banks, such as the World Bank and the African Development Bank, which provide loans and grants.

Contributions from donors, development banks, and (sub)regional organizations have declined for a large number of countries, which can be attributed largely to the overall decline in World Bank–funded projects since the late-1990s and early 2000s. Such projects variously focused on agricultural and economic development, and in terms of agricultural research generally aimed to reshape national agricultural research systems, provide much-needed

Figure 7—Examples of funding diversity, various years

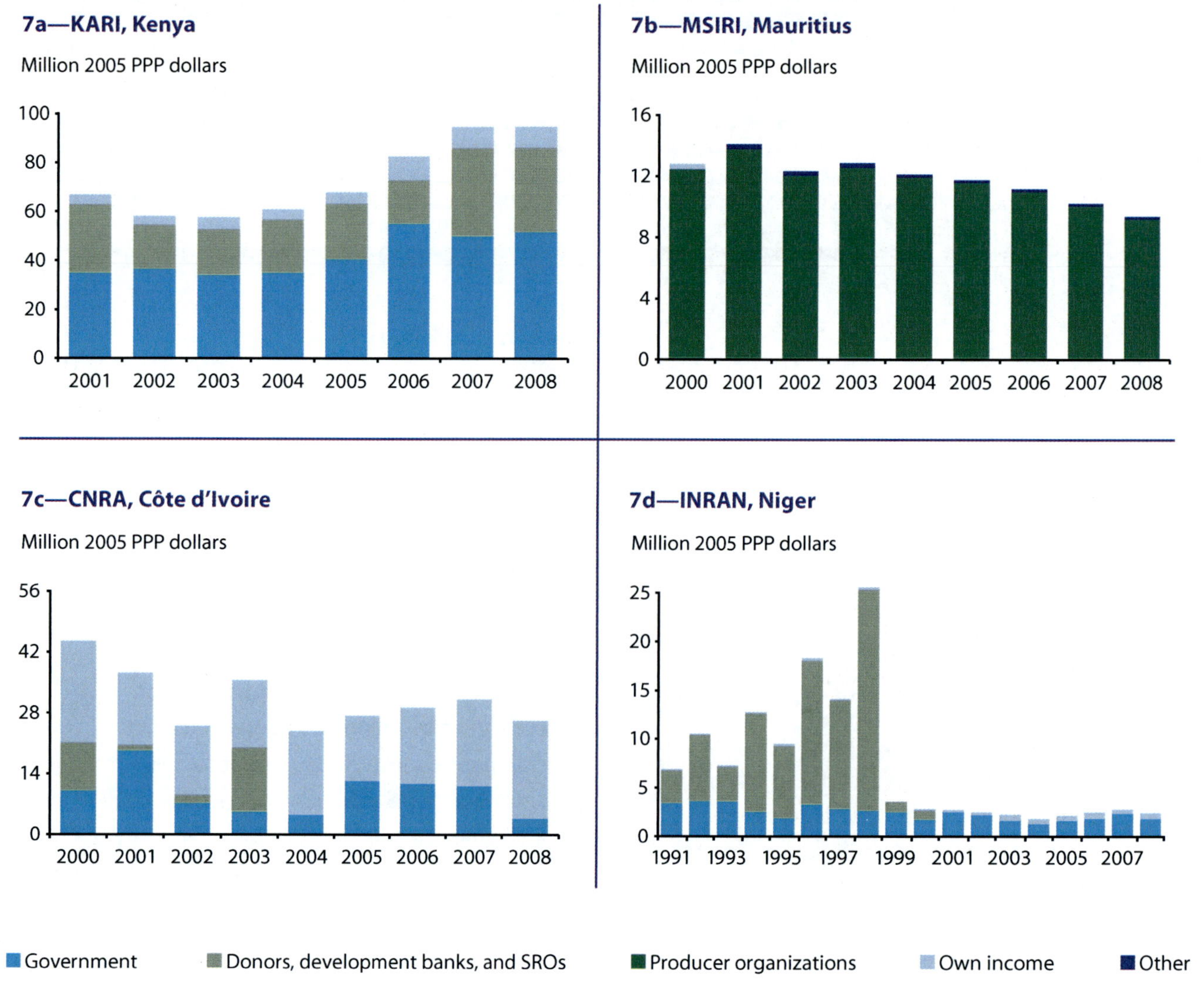

Source: Compiled by authors based on country-level ASTI survey data.

Notes: SROs indicates subregional organizations; "producer organizations" include contributions through export or production levies; "own income" includes sales of goods and services and contractual research performed for public and private agencies.

training opportunities, and support management and coordination initiatives (Beintema and Stads 2006). The completion of World Bank–supported projects led to a sharp funding decline in Niger and significant but less drastic declines in Guinea, Senegal, and Zambia. The highly unstable inflow of donor and development bank funding has negatively affected the institutional environment and research outputs in these countries, raising questions as to the long-term effectiveness of this type of funding. Some call for donors to support agricultural R&D over longer term periods (for example, 15 to 20 years), while at the same time expecting national governments to provide higher and more stable levels of funding to reduce future donor dependency. Donors, development banks, and African governments have to focus on the longer term in order to promote system stability, financial and institutional efficiency, and the overall quality of research outputs.

Although data are available only until 2008, contributions by donors and development banks are believed to have rebounded in more recent years with the launch of sizable projects funded through World Bank loans in a number of countries as part of the West Africa Agricultural Productivity Program (WAAPP) and the East Africa Agricultural Productivity Program (EAAPP). Activities focus on generating and disseminating improved agricultural technologies that address both national and regional priorities. To date WAAPP has initiated activities focusing on roots and tubers in Ghana, rice in Mali, and cereals in Senegal, and from 2011 Phase Two will focus on fruits and vegetables in Burkina Faso, livestock breeding in Niger, and further activities in another seven countries. Similarly, as of 2009–2010 EAAPP is funding research focusing on cassava in Uganda, rice in Tanzania, wheat in Ethiopia, and dairy in Kenya (World Bank 2007b and 2009).

Private Sector. Commercializing research outputs is often achieved through partnerships with the private sector.[9] In a few of the region's countries, private investment in agricultural R&D is increasing and creating an income stream for agricultural research agencies (Echeverría and Beintema 2009). In Senegal, for instance, large companies like SODEFITEX (cotton producers) and SUNEOR (groundnut producers) fund research activities conducted by the Senegalese Agricultural Research Institute (ISRA), the country's main government agricultural R&D agency (Stads and Sène 2010). These activities are often ad hoc, such as

Box 3—Private-sector funding in Côte d'Ivoire

CNRA's funding structure constitutes a unique and exemplary regional case study. The second National Agricultural Services Support Project (PNASA II), which was launched in 1998 and administered by the World Bank, stipulated that CNRA would be structured as a public–private entity, with 40 percent of its funding being contributed by the government and 60 percent derived from the private sector. To this end, the Inter-Professional Fund for Agricultural Research and Extension (FIRCA) was established in 2002. FIRCA relies on financial contributions not only from the government but also from the country's producers, who pay membership subscription dues through commodity-specific producer organizations. At least 75 percent of the subscription fees raised through agricultural production in a given subsector are allocated to programs serving the needs of that subsector. The remaining funds are allocated to a solidarity fund, and a marginal share underwrites FIRCA's operating costs. The purpose of the solidarity fund is to finance programs designed to serve production sectors (mostly food crops) unable to raise sufficient funding through their own subscription fees or that have difficulty doing so because of the way they are structured. In 2008, the amounts raised and contributed by the coffee, cocoa, rubber, and oil palm producer organizations represented 91 percent of total subscription dues raised by all the producer organizations combined.

Source: Stads and Doumbia 2010.

research on cash crops at ISRA in Senegal, but in many countries they are formalized, for example at CNRA in Côte d'Ivoire (Box 3). Although private-sector funding offers valuable potential support in developing financial and human resource capacity in agricultural R&D, it can be implemented only in countries with the necessary enabling policy environment, including strong intellectual property legislation, minimal barriers to importing and testing new technologies, and tax exemptions on research expenditures and venture capital (Alston, Pardey, and Piggott 2006). In many countries in SSA, the necessary policy environment is extremely weak or nonexistent. Furthermore, in many countries internally generated income is channeled back to the treasury, eliminating any incentive for research agencies to explore contract-based research for the private sector. It should be borne in mind that the proliferation of contracts to carry out research on behalf of agro-businesses entails the risk of excessively skewing the research agenda away from basic research in favor of applied research and seed multiplication. Attention to the balance between these types of research is therefore needed when private contract-based funding is expanded.

Commodity Levies. Research can also be funded through levies on agricultural production or exports. Commodity levies are important in several countries in SSA, including Kenya (coffee and tea), Malawi (tea and tobacco), Mauritius (sugar), South Africa (sugar), Tanzania (tea and coffee), and Zambia (cotton). Research levies have often been established in countries experiencing long-term funding instability that have an identifiable group of beneficiaries able to contribute to the cost of research. In other countries, levies date back to the colonial times. Levies are mostly applied to cash or export crops because they pass through a limited number of collection points. They are a less efficient mechanism for commodities and countries where most of the output is consumed on-farm or are traded in local markets because collection costs would be too high. Levies have the benefit of involving farmers in the research agenda and providing relevant research outputs. They may provide additional funding for agricultural R&D, but there is always the risk that they will simply replace rather than augment government funding.

Levies do, however, have a number of potentially negative impacts. These include the risks of promoting price disincentives and suboptimal funding levels due to spillovers and delayed research outputs. One way to counter these potential risks is for governments to provide funding to match the commodity levies (Kangasniemi 2002; Echeverría and Beintema 2009).

Competitive Funding Mechanisms. Competitive funding mechanisms have gained ground but are limited in Africa compared with other developing regions of the world. These funds typically finance R&D through grants allocated on the basis of scientific merit and congruence with broadly defined agricultural R&D priorities. Competitive funds are believed to attract research resources while lowering execution costs, encouraging demand-driven activities, and promoting research partnerships (Echeverría and Beintema 2009). A main concern, however, is long-term sustainability, given that many mechanisms are dependent on external funding. In the late 1990s and early 2000s, various competitive funds were established as components of World Bank projects in countries like Kenya, Mali, Senegal, and Tanzania. Though many funds have built-in sustainability mechanisms, overall funding will decline once initial endowments have been exhausted. Other countries, including Nigeria, Uganda, and Zambia, have attempted to establish competitive grant schemes, but without substantial initial injections of funding, these schemes have generally faltered.[10] Various regional and subregional competitive funds have also been established in recent years. The subregional organizations ASARECA and CORAF/WECARD both launched competitive schemes in the mid-2000s with a view to promoting demand-driven, collaborative research activities based on agreed criteria and priorities.

Staffing Trends

Agricultural Researcher Qualifications and Training

Thirty percent of the agricultural researchers employed in the ASTI countries in 2008 were qualified to the PhD level, 43 percent to the MSc level, and 27 percent to the BSc level. Researcher qualifications varied considerably across countries and by gender (Box 4 and Figure 8). In just 14 of the 32 ASTI countries, more than 80 percent of the FTE researchers were trained to the postgraduate (PhD or MSc) level. More than half the FTE researchers employed in Burkina Faso, Côte d'Ivoire, and Senegal and more than 40 percent of those employed in Benin, the Republic of Congo, and South Africa were trained to the PhD level. It is striking that many West African countries have maintained relatively large pools of well-qualified researchers despite recent losses in human and financial resource capacity. These high shares stem in large part from training programs conducted during the 1990s (and earlier), funded through bilateral donors or World Bank–financed projects.

South Africa's universities predate those of the rest of the region, which explains the consistently high number of well-trained agricultural research staff in that country. Moreover, a high and growing number of students from other SSA countries are attending South African universities, increasing the importance of their role in training future generations of African agricultural R&D staff. With two-thirds or more of the agricultural research staff qualified to the BSc level only, researchers in Eritrea and Guinea are the least

Box 4—Female participation in agricultural research

Women's participation in agricultural research has increased around the world over the past several decades, but it remains low in many countries, especially in the developing world. In 2008, 22 percent of FTE researchers in the ASTI countries were female, compared with 18 percent in 2001. Shares of female scientists in East and southern Africa are generally higher than those in West Africa. More than 30 percent of the agricultural researchers in Botswana, Eritrea, Mauritius, South Africa, and Sudan in 2008 were female. In contrast, corresponding shares in Ethiopia, Guinea, Mauritania, Niger, Senegal, and Sierra Leone were less than 10 percent. Overall, women are more represented in junior roles requiring only a BSc-level qualification. Female researchers also tend to be more prevalent in the higher education sector. Although the share of professional women employed in agriculture is increasing, the vast majority are entry-level staff and students (that is, BSc graduates or students). The need for greater representation by women in agricultural research in SSA is urgent. Women in senior positions as scientists, research managers, lecturers, and professors can provide valuable insights and perspectives to assist research agencies in addressing the unique and pressing challenges facing African farmers, many of whom are women.

Source: Country-level ASTI survey data and Beintema and Di Marcantonio 2010.

Figure 8—Distribution of agricultural researchers by country and degree qualification, 2008

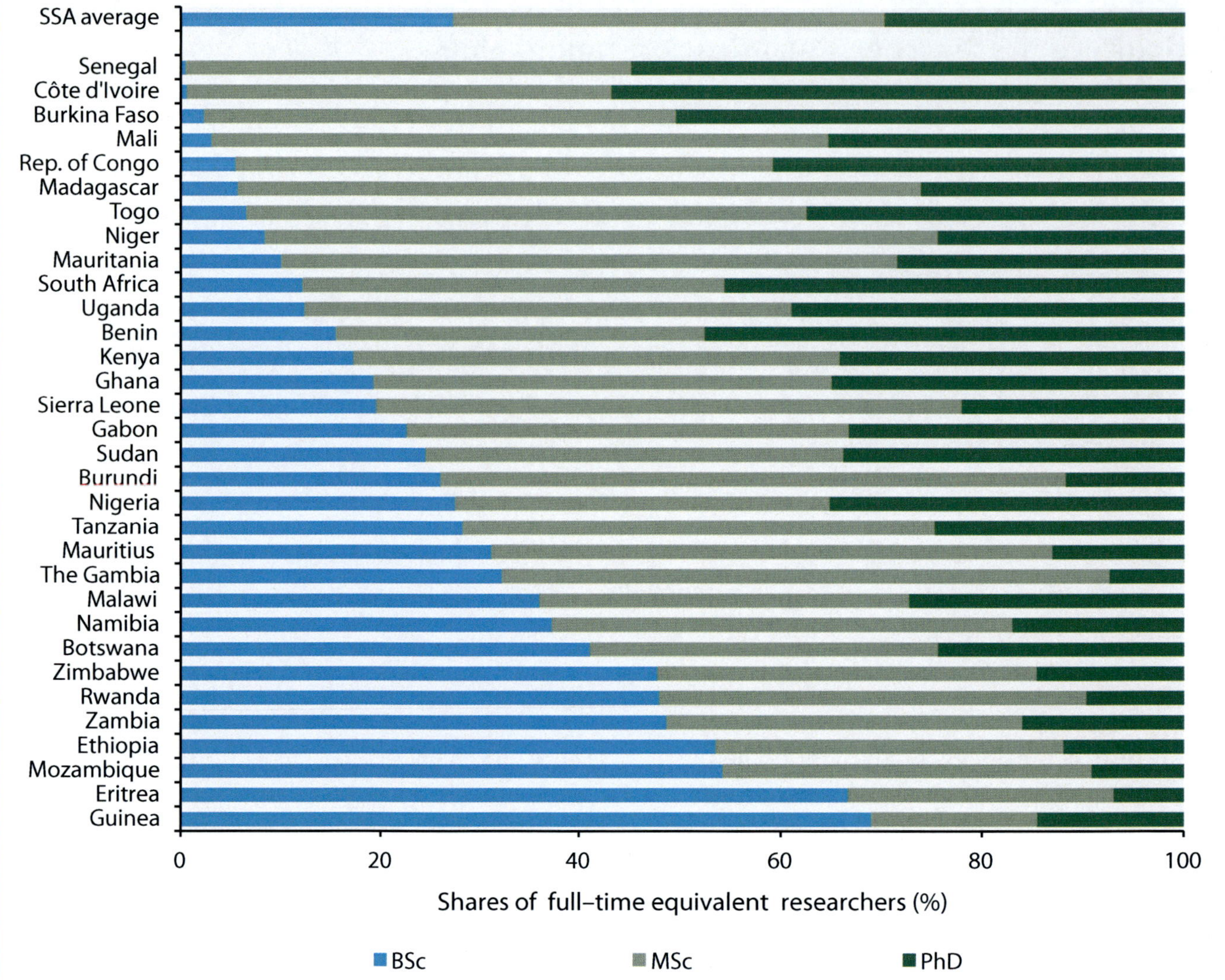

Source: Compiled by authors based on country-level ASTI survey data.

qualified of those in the ASTI countries. BSc-qualified researchers accounted for 54 percent of agricultural FTEs in Ethiopia and Mozambique, which is also very high. The overall limited research capacity in small countries, alongside low numbers of well-qualified staff and limited training opportunities, poses significant constraints on their ability to conduct high-quality research and attract external funding. As an example, The Gambia employed only 38 FTE agricultural researchers in 2008, just two of whom were trained to the PhD level (Stads and Manneh 2010).

Despite growth in agricultural R&D capacity across the region, average levels of staff qualifications actually deteriorated somewhat in a combined sample of 30 ASTI countries (excluding Mozambique and Rwanda). In 2008, 27 percent of all FTE researchers held BSc degrees, up from 24 percent in 2001 (Figure 9). The overall share of PhD-qualified researchers rose only slightly during 2001–2008, from 29 to 30 percent, but trends were more pronounced in certain countries. In 2008, half of Zambia's public agricultural researchers were trained to the MSc and PhD levels—a significant shift from 2001, when 70 percent of researchers held postgraduate degrees. The increasing share of BSc-qualified staff in Zambia stems from a government-sector hiring freeze, lack of appropriate training opportunities, and the concurrent reduction in the number of senior researchers, mainly based on retirement and losses to

Figure 9—Distribution of agricultural researchers by degree and institutional category, 2001 and 2008

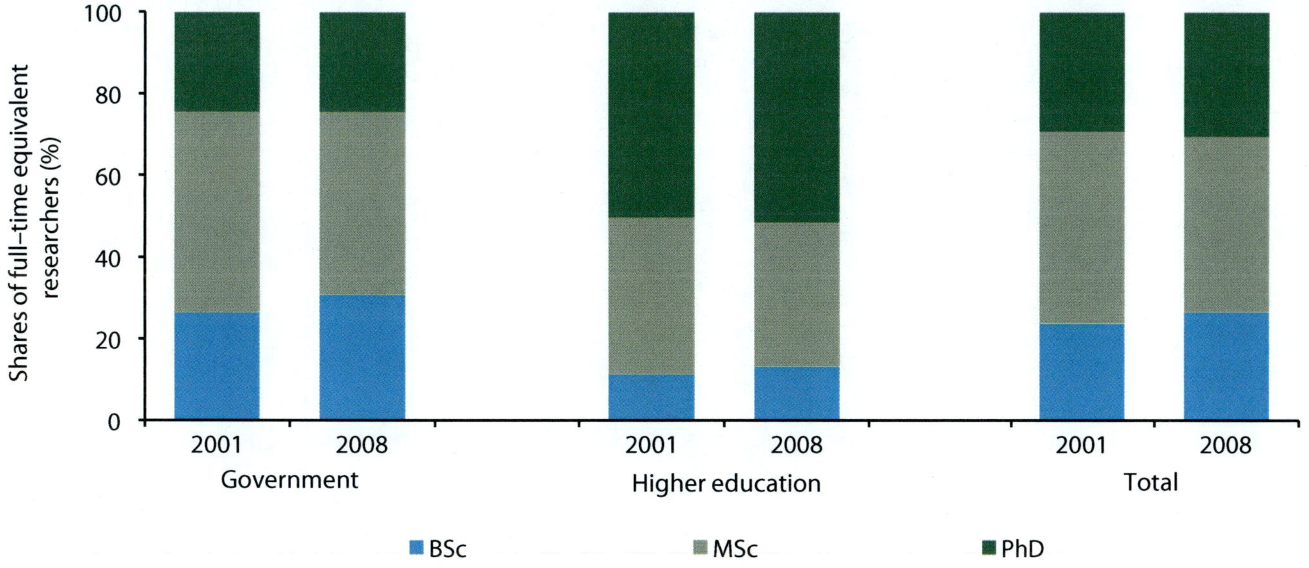

Changes in shares of BSc-qualified staff:

- 3–4 percent growth in Ethiopia, Ghana, and Uganda
- 6 percent growth in Nigeria
- 8 percent growth in Botswana
- 20 percent growth in Zambia
- Constant in many other countries

Source: Compiled by authors based on country-level ASTI survey data.

Notes: Data exclude Mozambique and Rwanda because 2001 data were not available. "Total" includes nonprofit institutions.

other agencies (Flaherty and Mwala 2010). In Nigeria, the overall share of PhD- and MSc-qualified researchers fell from 79 to 72 percent during 2001–2008, and in Botswana this share fell from 66 to 59 percent. In Ethiopia, Ghana, and Uganda postgraduate shares declined by 3 to 4 percentage points during this time. Higher shares of BSc-qualified researchers were often the result of recruitment bans being lifted, lack of training opportunities, and declining numbers of senior staff.

The share of PhD-qualified staff was higher in the higher education sector than in the government or nonprofit sectors—52 percent in 2008 compared with 24 and 29 percent, respectively—but this is common in most SSA countries and others around the world. The deterioration in average qualification levels was more pronounced at government agencies, where the share of BSc-qualified researchers increased from 27 to 31 percent during 2001–2008 compared with 11 to 13 percent at the higher education agencies.

During the 1970s and 1980s, many countries received considerable financial support for staff training, often as part of large World Bank–financed projects or through contributions from bilateral donors. By the late 1990s, most donors had either cut or eliminated their funding for graduate training (Beintema and Stads 2006).[11] The new World Bank projects previously described (WAAPP and EAAPP) have brought new opportunities, as have other projects supported by the European Union and other donors. The Alliance for a Green Revolution in Africa (AGRA) is another initiative offering training opportunities. The Alliance has provided (or will shortly provide) funding for 80 PhD fellowships in plant breeding and 170 MSc fellowships in agronomy. AGRA also plans to strengthen crop

Figure 10—University-qualified research support staff as a share of total research staffing, selected countries, 2008

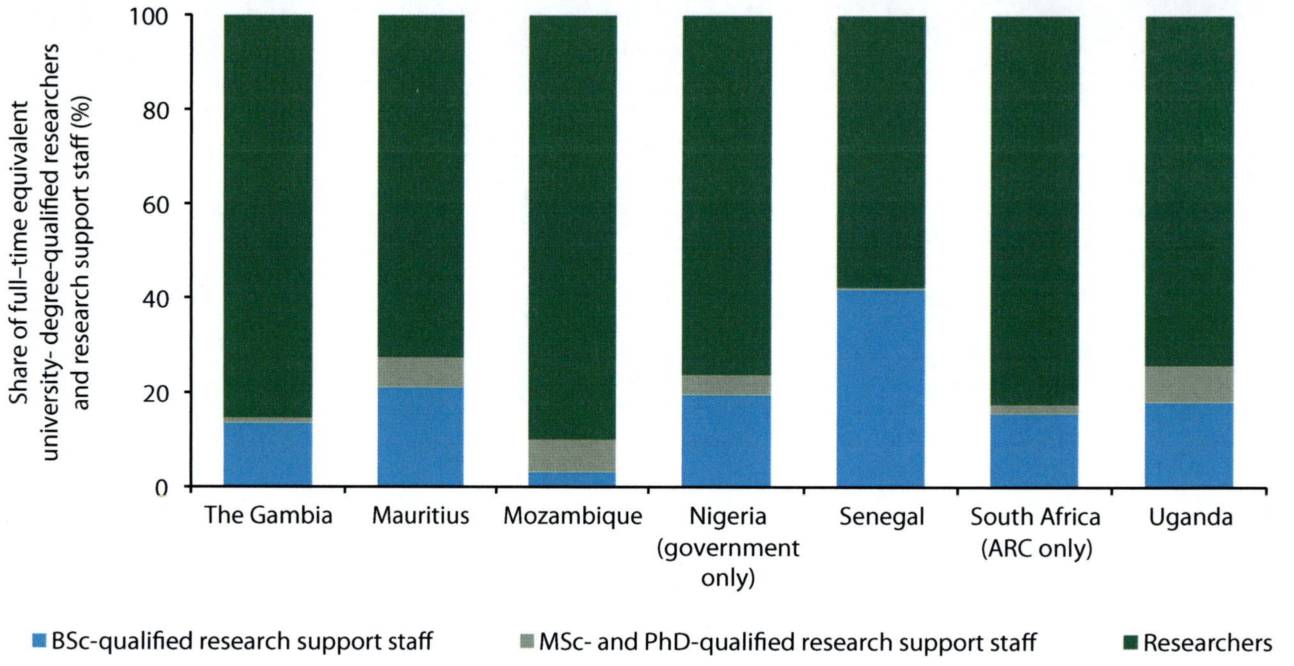

Source: Compiled by authors based on country-level ASTI survey data.

science programs in at least 10 African universities by providing training in soil health disciplines for 40 to 50 PhD scientists, 120 MSc scientists, and 200 laboratory technicians (AGRA 2010).

An increasing number of support staff (technicians, research assistants, laboratory assistants) have BSc, MSc, and occasionally PhD qualifications, but they are not classified as researchers (Figure 10). In Senegal, for example, the minimum requirement for a researcher is an MSc degree, so the 105 scientists employed at ISRA with BSc qualifications are classified as technicians (Stads and Sène 2010). Half of the technicians and other research support staff at NARO in Uganda held MSc or BSc degrees, and most attained these qualifications without official NARO backing. Although the number of research positions at NARO increased in recent years, promotional opportunities remain limited because applicants must meet specific minimum requirements, including having an MSc degree (Flaherty, Kitone, and Beintema 2010). Unlike the situations in Senegal and Uganda, support staff at Tanzania's DRD are promoted to researcher status upon obtaining their BSc degrees (Flaherty and Lwezaura 2010). The pool of degree-qualified support staff is sizeable in some countries. In Senegal, 43 percent of all degree-qualified research staff are technicians, and in Mauritius, Nigeria (government agencies only), and Uganda this share is about 25 percent (Figure 10). It is important to capture quantitative information on research technicians, who—given proper training and promotional opportunities—will be a valuable resource in the future development of African agricultural R&D.

Replacing an Aging Pool of Senior Researchers

In many of the region's countries, salary and retirement packages and conditions of service are poor. In addition, many agencies have outdated infrastructure and insufficient operating budgets to conduct research. Even with the aforementioned increase in training opportunities in a number of countries, research agencies have difficulty retaining staff once they attain higher degrees and can attract offers of better remuneration and conditions, either in the higher education or private sectors or abroad (FARA 2006; World Bank 2007c). A major concern in many countries, particularly in West and Central Africa, is a rapidly aging

Figure 11—Age distribution of agricultural researchers at the main government and higher education agencies, selected countries, 2007

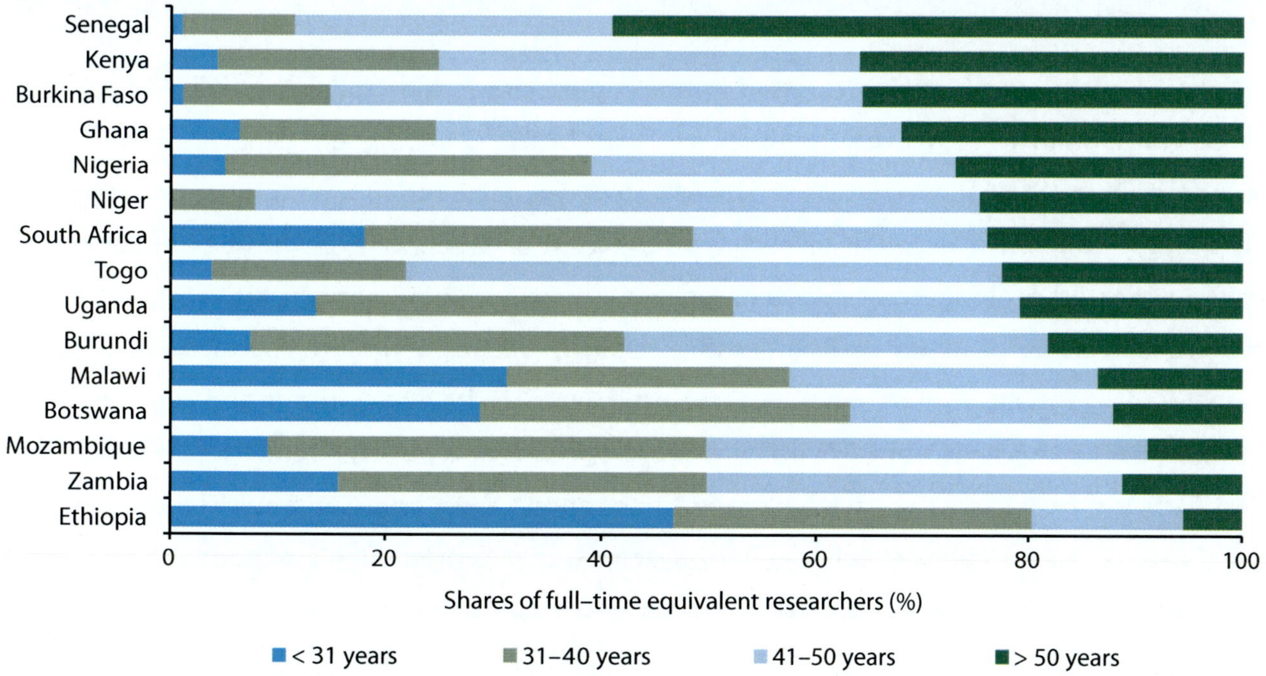

Source: Compiled by authors based on ASTI–AWARD 2008.

pool of scientists, many of whom will approach retirement within the next decade. In Cameroon and the Republic of Congo, for instance, agricultural researchers at the main agencies are well over 50 years old on average. In 2007, 27 percent of researchers in Ghana's CSIR were 51 years or older, and half were between 41 and 50 years old (Figure 11).[12] That year, 36 percent of KARI's researchers in Kenya were 51 years or older. With 59 percent of its researchers averaging well over 50 years in 2007, Senegal has one of the oldest pools of scientists in West Africa. In 2003, ISRA in Senegal employed 70 PhD-qualified scientists compared with just 54 in 2008. Some of these scientists left ISRA to take advantage of opportunities in the higher education and private sectors, where salaries are reported to be up to three times higher than in the public sector; many of the more senior researchers also retired (Stads and Sène 2010).

A number of agencies are instituting staff retention strategies. KARI in Kenya, for example, has introduced regular staff performance evaluations, which form the basis for promotion. The institute is also working on other incentives, such as better medical benefits. KARI also requires that staff offered training commit to working for the agency for a set period of time. In the mid-2000s, KARI and other institutes convinced the government to increase the retirement age from 55 to 65 years, not only to address the shortage of senior staff, but also to offset the time it takes for staff to qualify for and then undertake MSc and PhD training. It made sense to extend the productivity of these researchers once they became fully qualified. The higher retirement age also provided an incentive for junior staff, including technicians with diplomas, to pursue higher training, even through self-sponsorship (Flaherty et al. 2010b).

Attracting and retaining staff is an even more seriously problem in countries with small research capacities. The National Agricultural Research Institute in The Gambia is a case in point. During 2003–2009, the institute lost seven PhD-qualified researchers through retirement, departure, or death, and many of its remaining staff members lack advanced training or experience (Stads and Manneh 2010). This lack of a critical mass of well-qualified researchers in small countries also highlights the need for regional initiatives focusing on the needs and vulnerabilities of such countries.

Future Directions to Address Current Challenges

New quantitative evidence presented in this report shows that, following a period of stagnation in the 1990s, total public agricultural R&D spending and capacity levels in SSA have increased. Most of the investment growth, however, occurred in a handful of countries; in many other countries, investment levels have stagnated or fallen. A large number of countries variously reported prolonged recruitment freezes, limited training opportunities, aging pools of researchers, losses of senior staff, and, more recently, disproportionate recruitment of junior, BSc-qualified scientists. Some countries currently have such low investment and capacity levels that the impact of agricultural R&D on rural development and poverty reduction is questionable at best. This is particularly the case for many francophone West and Central African countries that have extremely fragile agricultural R&D systems, that remain highly dependent on external funding, and whose agricultural researchers are rapidly approaching retirement age.

Well-developed national agricultural research systems and adequate levels of investment are important prerequisites for agricultural development, food security, and poverty reduction. In recent years, governments have exhibited renewed interest in supporting agricultural development in SSA. CAADP, the G8 L'Aquila summit, the UN High-Level Task Force on the Global Food Security Crisis, as well as international efforts to re-engage in climate change mitigation and natural resources management all contributed to returning agriculture and agricultural R&D to the political agenda. But this political support must be translated into a set of specific directives by governments, donors, and other R&D stakeholders if the many challenges facing agricultural R&D systems are to be addressed. Various highly influential reports and meetings, including the *World Development Report 2008* (World Bank 2007a); the International Assessment of Agricultural Knowledge, Science and Technology for Development *Synthesis Report* (IAASTD 2008); and the Global Author Team report (Lele et al. 2010) for the Global Conference on Agricultural Research for Development have provided a framework of policy recommendations. In particular, the InterAcademy Council (IAC) report *Realizing the Promise and Potential of African Agriculture* (IAC 2004) provides a comprehensive list of strategic recommendations supported by detailed actions for different target audiences (see Chapter 8 and Annex B of the report). Most of the policy directives and actions recommended by IAC and others still apply today, despite significant regionwide investment and capacity growth in agricultural R&D since the turn of the millennium. Building on the strategic recommendations of these initiatives and taking into account the various investment and capacity challenges outlined in this report, governments, donors, and other stakeholders must address four key areas of policy implications. These four areas are discussed in the following sections.

Counteracting Decades of Underinvestment in Agricultural R&D. Repeated calls have been made for increased investments in the agricultural research systems of developing countries in recent years (IAC 2004; World Bank 2007a; IAASTD 2008; Lele et al. 2010). As evidenced in this report, a number of national governments have stepped up their allocations to agricultural research, but overall investment levels in most SSA countries are still below the levels required to sustain agricultural R&D needs. Countries that have increased their expenditures have directed most of the funds toward salary increases and the rehabilitation of infrastructure and equipment. Nevertheless, these important investments have to be complemented with additional allocations to increase the number, variety, and intensity of actual research activities. National governments should urgently address underinvestment in agricultural R&D. They will need to make more funds available to support research activities if investments are to translate into improved agricultural productivity. Increased government support to agricultural R&D should also include funding to allow universities to establish and maintain basic research programs, which to date have been limited. Moreover, governments, donors, and regional and international organizations must cooperate more closely and increase their commitments to agricultural R&D if SSA countries are to meet CAADP's 6 percent yearly target for AgGDP growth or the poverty and hunger targets of the Millennium Development Goals. Finally, diversification of funding sources is needed, for example, through the sale of goods and services and increased participation in and funding of research by the private sector. As stated, this, in turn, requires that national governments provide a more enabling policy environment.

Halting Excessive Volatility in Yearly Investment Levels. The time-series data in this report reveal that agricultural R&D funding in SSA has been highly volatile. Many countries continue to be extremely dependent on unstable inflows of donor funding and development bank loans, and in many instances the completion of large donor-funded projects has precipitated severe financial crises, seriously undermining any progress made. In this way, very often the gains achieved through donor-funded projects are quickly eroded in the absence of viable mechanisms to sustain them. Donor funding is typically short term and ad hoc, calling into question the long-term effectiveness and efficiency of this type of funding. In addition, research by nature involves inherent time lags between investment in R&D and the attainment of returns to those investments in the form of tangible benefits (Alston, Pardey, and Piggott 2006); this further highlights the need for long-term, stable funding. Volatility in year-to-year spending levels can be halted only through stable and sustainable levels of government funding. Governments have to clearly identify their long-term national R&D priorities and design relevant, focused, and coherent R&D programs accordingly. Donor funding needs to be better aligned with national priorities, and consistency and complementarities between donor programs need to be ensured. Real progress can be achieved only with sustained, long-term backing from national governments, donors, and regional and international organizations.

Addressing Existing and Imminent Challenges in Human Resource Capacity. Growing concern exists regarding the lack of human resource capacity in agricultural R&D to enable satisfactory responses to emerging global challenges. National governments and donor organizations must expand their investments in agricultural higher education to allow universities to increase the number and size of their MSc and PhD programs and to improve the curricula of existing programs. The regional community has an important role to play in this regard, particularly when it comes to small countries with limited or nonexistent MSc or PhD training opportunities. In recent years, various regional capacity-building initiatives have begun, but these will have to be further expanded in order to address some of the capacity challenges evidenced in this report, including aging pools of scientists and increasing shares of junior research staff in a large number of countries. As a result of prolonged recruitment freezes, many countries lack middle-level staff needed both to take on seniority as older scientists retire and to train and mentor the younger researchers coming up behind them. In addition to university training programs, agricultural research agencies will need to establish mentoring programs to facilitate on-the-job training for junior scientists. National governments must also promote (agricultural) science as a valuable career path for young people, which should include strengthening primary- and secondary-level education in the sciences. Moreover, many countries with serious capacity gaps will have to increase the civil servant retirement age or institute flexible working arrangements to ensure that retired researchers can contribute to much-needed training and mentorship initiatives.

Maximizing Regional and Subregional Cooperation in Agricultural R&D. Because of the high fixed costs inherent in research, small countries generally lack the required critical mass of agricultural R&D capacity and hence face enormous challenges in producing or accessing relevant, high-quality research outputs (World Bank 2007a). Very often, the only viable—and efficient—solution is regional collaboration. Through regional initiatives, technological innovation in one country can quickly have an impact in other countries with similar agroclimatic conditions, creating what is known as a leapfrog effect. Creative efforts to build and enhance strong subregional linkages need to be further strengthened in order to maximize these synergistic opportunities. Because many of the regional efforts have a network approach, the CGIAR will continue to act as a critical provider of agricultural technologies in most SSA countries, as well as supporting capacity building efforts.

Monitoring the performance, inputs, and outcomes of agricultural S&T systems is fundamental to assessing progress toward CAADP's targets and the strategic recommendations and policy directives espoused in the various influential reports and meetings described above. Up-to-date information is critical to accurate interpretations of the current status and direction of national agricultural research systems in SSA countries. Regular collection of data on agricultural S&T capacity and investments, as undertaken by ASTI, is therefore essential.

Notes

1. The excluded countries (a) have very small agricultural research capacity (for example, Cape Verde and Swaziland), (b) did not respond to our request to collaborate (for example, Cameroon and Chad), and/or (c) were unable to provide sufficient information (for example, Angola and Democratic Republic of Congo).

2. Note that this analysis will be complemented by a series of reports to be published in 2011 that provide more in-depth analyses of the trends described here.

3. See Roseboom, Pardey, and Beintema (1998) and Beintema and Stads (2006) for a detailed history of institutional developments in agricultural R&D in SSA.

4. ASTI is involved in a separate survey on the role of the private sector in agricultural R&D and innovation in Kenya, Senegal, South Africa, Tanzania, and Zambia. The results of this study are currently being synthesized and will be available later in 2011.

5. SADC is in the process of establishing the Centre for Coordination of Agricultural Research and Development in Southern Africa (CCARDESA), along the lines of ASARECA and CORAF/WECARD. CCARDESA will become operational in 2011.

6. This report focuses only on the number of researchers per million economically active population; other intensity ratios are accessible via ASTI's Data Tool at www.asti.cgiar.org/data.

7. Note that this total does not include R&D spending by regional and international organizations, such as the research centers of the CGIAR.

8. This is a reference to what is known as Africa's Big Five—the lion, leopard, rhinoceros, elephant, and buffalo.

9. These partnerships have the potential to accelerate the development of new technologies and reduce the costs of research. However, they apply only to technologies for which benefits can be appropriated (Spielman, Hartwich, and von Grebmer 2007; Echeverría and Beintema 2009).

10. For example, although the competitive grant scheme for agricultural research in Nigeria was launched some time ago, the first set of applicants has yet to be announced (Flaherty et al. 2010a). Similarly, Zambia's Science and Technology Development Fund was enacted in 1997 but has yet to become operational (Flaherty and Mwala 2010).

11. Reasons for this include the high costs of training students overseas and the relatively low rate of students that returned after their studies (Beintema and Stads 2006).

12. During 2008, ASTI and the African Women in Agricultural Research and Development (AWARD) program conducted a survey to obtain gender-disaggregated capacity indicators for 125 agricultural research and higher education agencies in 15 SSA countries. See Beintema and Di Marcantonio (2010) for more information.

References

AGRA (Alliance for a Green Revolution in Africa). 2010. "AGRA's Programs." Accessed November 17, 2010. http://www.agra-alliance.org/section/work.

Alston, J. M., P. G. Pardey, and R. R. Piggott. 2006. "Synthesis of Themes and Policy Issues." In *Agricultural R&D in the Developing World: Too little, Too Late?*, edited by J. M. Alston, P. G. Pardey, and R. R. Piggott. Washington, DC: International Food Policy Research Institute.

ASTI (Agricultural Science and Technology Indicators). ASTI website. http://www.asti.cgiar.org.

ASTI–AWARD (Agricultural Science and Technology Indicators and African Women in Agricultural Research and Development). "2008. Gender-Disaggregated Capacity Indicators in Sub-Saharan Africa, 2007/08." Unpublished surveys. International Food Policy Research Institute, Washington, DC, and CGIAR Gender and Diversity Program, Nairobi.

Beintema, N. M., and F. Di Marcantonio. 2010. *Female Participation in African Agricultural Research and Higher Education: New Insights. Synthesis of the ASTI–Award Benchmarking Survey on Gender-Disaggregated Capacity Indicators.* IFPRI Discussion Paper 957. Washington, DC, and Nairobi: International Food Policy Research Institute and African Women in Agricultural Research and Development.

Beintema, N. M., and G. J. Stads. 2006. *Agricultural R&D in Sub-Saharan Africa: An Era of Stagnation.* ASTI Background Report. Washington, DC: International Food Policy Research Institute.

———. 2008a. *Measuring Agricultural Research Investments: A Revised Global Picture.* ASTI Background Note. Washington, DC: International Food Policy Research Institute.

———. 2008b. *Diversity in Agricultural Research Resources in the Asia-Pacific Region.* Washington, DC, and Bangkok: International Food Policy Research Institute and Asia-Pacific Association of Agricultural Research Institutions.

———. 2010. *Public Agricultural R&D Investments and Capacities in Developing Countries: Recent Evidence for 2000 and Beyond.* Background note prepared for the Global Conference on Agricultural Research for Development, Montpellier, France, March 28–31. Washington, DC: International Food Policy Research Institute.

———. Forthcoming. *ASTI Toolkit—Monitoring Agricultural R&D Capacity and Investment Indicators: A Practitioner's Guide to ASTI's Methodologies & Data Collection Standards.* Washington, DC: International Food Policy Research Institute.

Echeverría, R. G., and N. M. Beintema. 2009. *Mobilizing Financial Resources for Agricultural Research in Developing Countries: Trends and Mechanisms.* Rome: Global Forum on Agricultural Research.

FARA (Forum for Agricultural Research in Africa). 2006. *Agricultural Research Delivery in Africa: An Assessment of the Requirements for Efficient, Effective and Productive National Agricultural Research Systems in Africa: Main Report and Strategic Recommendations.* Accra, Ghana: FARA.

FAO (Food and Agriculture Organization of the United Nations). 2009. FAOSTAT Database. Accessed March 23, 2010. http://faostat.fao.org/site/452/default.aspx.

Flaherty, K., and D. Lwezaura. 2010. *Tanzania.* ASTI Country Note. Washington, DC, and Dar es Salaam: International Food Policy Research Institute and Department of Research and Development.

Flaherty, K., and M. Mwala. 2010. *Zambia.* ASTI Country Note. Washington, DC, and Lusaka: International Food Policy Research Institute.

Flaherty, K., G. O. Essegbey, and R. Asare. 2010. *Ghana.* ASTI Country Note. Washington, DC, and Accra: International Food Policy Research Institute and Council for Scientific and Industrial Research.

Flaherty, K., F. Kelemework, and K. Kelemu. 2010. *Ethiopia.* ASTI Country Note. Washington, DC, and Addis Ababa: International Food Policy Research Institute and Ethiopian Institute of Agricultural Research.

Flaherty, K., D. Kitone, and N. M. Beintema. 2010. *Uganda.* ASTI Country Note. Washington, DC, and Entebbe: International Food Policy Research Institute and National Agricultural Research Organisation.

Flaherty, K., G. F. Liebenberg, and J. Kirsten. 2010. *South Africa.* ASTI Country Note. Washington, DC, and Pretoria: International Food Policy Research Institute, Agricultural Research Council, and University of Pretoria.

Flaherty, K., F. Mazuze, and R. Mahanzule. 2010. *Mozambique.* ASTI Country Note. Washington, DC, and Maputo: International Food Policy Research Institute and Agricultural Research Institute of Mozambique.

Flaherty, K., G. Ayoola, J. Ogbodo, and N. M. Beintema. 2010a. *Nigeria.* ASTI Country Note. Washington, DC, and Abuja: International Food Policy Research Institute, Agricultural Research Council of Nigeria, and Farm and Infrastructure Foundation.

Flaherty, K., F. Murithi, W. Mulinge, and E. Njuguna. 2010b. *Kenya.* ASTI Country Note. Washington, DC, and Nairobi: International Food Policy Research Institute and Kenya Agricultural Research Institute.

IAASTD (International Assessment of Agricultural Knowledge, Science and Technology for Development). 2008. *Synthesis Report.* Washington, DC: Island Press.

IAC (InterAcademy Council). 2004. *Realizing the Promise and Potential of African Agriculture: Science and Technology Strategies for Improving Agricultural Productivity and Food Security in Africa.* Amsterdam: IAC.

Kangasniemi, J. 2002. "Financing Agricultural Research by Producers' Organizations in Africa." In *Agricultural Research Policy in an Era of Privatization*, edited by D. Byerlee and R. G. Echeverría. Wallingford, UK: CAB International.

Lele, U., J. Pretty, E. Terry, and E. Trigo. 2010. "Transforming Agricultural Research for Development." Report prepared for the Global Conference on Agricultural Research for Development, Montpellier, France, March 28–31. Rome: Global Forum on Agricultural Research.

Montpellier Panel. 2010. *Africa and Europe: Partnerships for Agricultural Development.* London: Imperial College.

NEPAD (New Partnership for Africa's Development, Office of Science and Technology). 2006. *Africa's Science and Technology Consolidated Plan of Action.* Pretoria: NEPAD.

NEPAD–CAADP (New Partnership for Africa's Development–Comprehensive Africa Agriculture Development Programme). 2010. "Pillar 4: Agricultural Research." Accessed November 2010. http://www.nepad-caadp.net/pillar-4.php.

Rahija, M., J. Ramkissoon, and G. J. Stads. 2010. *Mauritius.* ASTI Country Note. Washington, DC, and Port Louis: International Food Policy Research Institute and Food and Agricultural Research Council.

Roseboom, J., P. G. Pardey, and N. M. Beintema. 1998. *The Changing Organizational Basis of African Agricultural Research.* Environment and Production Technology Division Discussion Paper 37. Washington, DC: International Food Policy Research Institute.

Spielman, D. J., F. Hartwich, and K. von Grebmer. 2007. *Sharing Science, Building Bridges, and Enhancing Impact: Public–Private Partnerships in the CGIAR.* International Food Policy Research Institute (IFPRI) Discussion Paper 708. Washington, DC: IFPRI.

Stads, G. J., and N. M. Beintema. 2009. *Public Agricultural Research in Latin America and the Caribbean: Investment and Capacity Trends.* ASTI Synthesis Report. Washington, DC: International Food Policy Research Institute and Inter-American Development Bank.

Stads, G. J., and S. Doumbia. 2010. *Côte d'Ivoire.* ASTI Country Note. Washington, DC, and Abidjan: International Food Policy Research Institute and National Center for Agricultural Research.

Stads, G. J., and K. El-Siddig. 2010. *Sudan.* ASTI Country Note. Washington, DC, and Wad Madani: International Food Policy Research Institute and Agricultural Research Corporation.

Stads, G. J., and F. Manneh. 2010. *The Gambia.* ASTI Country Note. Washington, DC, and Banjul: International Food Policy Research Institute and National Agricultural Research Institute.

Stads, G. J., and L. Sène. 2010. *Senegal.* ASTI Country Note. Washington, DC, and Dakar: International Food Policy Research Institute and Senegalese Agricultural Research Institute.

Stads, G. J., M. Issoufou, and A. M. Massou. 2010. *Niger.* ASTI Country Note. Washington, DC, and Niamey: International Food Policy Research Institute and Niger National Institute of Agricultural Research.

World Bank. 2007a. *World Development Report 2008: Agriculture for Development.* Washington, DC.

———. 2007b. *First Phase of a Support Program to the West Africa Agricultural Productivity Program (WAAPP).* Project information document. Report 43482. Washington, DC.

———. 2007c. *Cultivating Knowledge and Skills to Grow African Agriculture.* Washington, DC.

———. 2009. *Project Appraisal Document on Three Proposed Credits in the Amount of SDR 20.1 Million (US$30 Million Equivalent) to the Federal Democratic Republic of Ethiopia, of SDR 20.1 Million (US$30 Million Equivalent) to the Republic of Kenya, of SDR 20.1 Million (US$30 Million Equivalent) to the United Republic of Tanzania for a Total of SDR 60.3 Million (US$90 Million Equivalent) for Eastern Africa Agricultural Productivity Program APL1 in Support of the First Phase of the Eastern Africa Agricultural Productivity Program.* Report 48295-AFR. Washington, DC.

———. 2010. *World Development Indicators.* Washington, DC. Accessed November 2010. http://data.worldbank.org/data-catalog/world-development-indicators/wdi-2010.

About the Authors

Nienke Beintema is program head of the International Food Policy Research Institute's (IFPRI) Agricultural Science and Technology Indicators (ASTI) initiative.

Gert-Jan Stads is program coordinator of IFPRI's ASTI initiative.

Made in the USA
Las Vegas, NV
12 May 2026

46757714R00026